D1357636

TH

NAVIGATION MANUAL

THE ULTIMATE
NAVIGATION
MANUAL

Lyle F. Brotherton

Collins

HarperCollins Publishers
77–85 Fulham Palace Road
London W6 8JB
www.harpercollins.co.uk

Collins is a registered trademark of HarperCollins Publishers Ltd.

First published 2011

A catalogue record for this book is available from the British Library.

ISBN 978-0-00-742460-3

Edited, illustrated and designed by Tom Cabot/ketchup

Printed and bound in China by South China Printing Company Ltd.

Picture Credits

All photos courtesy of the author, except: p. 5, John Cleare/Mountain Camera Picture Library; pp. 46, 145, 189, 190 & 194, Dennis Varouxis; p. 67, NOAA/NGDC/CIRES; pp. 76, 225 & 228, USAF; p. 77, Corbis/Paul A. Souders; pp. 128, 137, 138 & 155, USGS; p. 147, Corbis/George Steinmetz; p. 204, Corbis/Seth Resnick; p. 209, Corbis/Shuji Kotoh; p. 215, Corbis/Andrew McConnell; p. 224, Corbis/Steve Parish Publishing; all Ordnance Survey imagery © Crown copyright 2011; Anquet screengrabs © Anquet Technology Ltd; TopoPointUSA screengrabs © Ebranta Technologies Inc.

PREFACE

by Sir Ranulph Fiennes

Any venture into the outdoors carries with it a risk, from a day's walk in the hills to a polar expedition, and good navigation is the foundation of all safe adventure. Navigational errors are the most common contributory cause of mountain accidents,* often being the first link in a chain of events that can lead to catastrophe.

Search and rescue teams are regularly called out in the worst weathers and need to reach inaccessible areas quickly and safely. These same requirements apply to anybody who wants to safely enjoy the outdoors. This field instruction manual brings together for the first time 'Best Practice' as used by these teams throughout the world and makes it accessible to everyone.

Planning and preparation underpin any successes I may have had. Make them yours!

---

* Scottish Mountaineering Incidents(1996–2005), *Research Digest*, No. 102.

CONTENTS

SECTION THREE: SPECIAL ENVIRONMENTS

SECTION FOUR: GLOBAL NAVIGATION SATELLITE SYSTEMS AND DIGITAL MAPPING

SECTION FIVE: APPENDICES

FOREWORD

by David 'Heavy' Whalley

Whether your interest lies in reaching one of the earth's most remote destinations, undertaking your first Duke of Edinburgh outbound challenge or simply enjoying a day's walk in the hills, true freedom and safety outdoors depends on proficient navigational skills.

Micronavigation is an extremely accurate system of land navigation for all terrains: Alpine, Arctic, Desert, Forest, Jungle, Mountain, Shoreline and Urban. The techniques can be mastered by anyone from a hiker to a search and rescue (SAR) responder, and are used to navigate competently and securely in every environment and in all conditions.

Encompassing the full range of navigational techniques available, and the thinking and technology which supports them – from the direction of the wind and simple compasses, to celestial direction-finding and cutting-edge, global satellite systems – this manual brings together for the first time all of these straightforward techniques.

The end result is one of the most comprehensive and easy-to-use navigation manuals ever made available. This book is a landmark in land navigation.

David 'Heavy' Whalley, BEM, MBE and the Distinguished Service Award for Service to Mountain Rescue. Team Leader of RAF Leuchars MRT (Mountain Rescue Team), RAF Kinloss MRT and Deputy Team Leader at RAF Valley MRT, he has taken part in more than 1,000 mountain rescue and air-crash searches including the Lockerbie Pan Am Flight 103 and has helped in saving hundreds, probably thousands, of lives.

ACKNOWLEDGEMENTS

Firstly, I would like to thank my wife, Judy, for allowing me to spend so much time away from Ramshead over the last six years visiting and working with search and rescue teams (SAR) and members of the Special Forces with whom I have had the privilege to navigate.

Their names would fill an entire book. Those who I have chosen to name are representative of the many hundreds of specialists I have met throughout the world and with whom I have been fortunate enough to make many new friends, learning something from each and every one of them: my profound thanks to you all.

The majority of the photography and art direction for this project has been undertaken by two very close friends and commensurate professionals to whom I am eternally grateful:

- Vaughan Judge – whose work has been included in numerous international exhibitions and publications. In 2007, his photographic work was included in the first comprehensive history of Scottish photography – *Scottish Photography: A History*. Vaughan has taught photography at art schools across Europe and America.
- Kate Jo – who is a practising visual artist who has exhibited internationally and has been the recipient of various awards. Her work has been featured in publications such as *Next Level* and *A-N Magazine*. Kate works as a freelance photographer and videographer and is currently the Gallery Director for Montana State University School of Art.

And the global experts:

- Stuart Johntson MIC, WEMT. Team Leader Tayside MRT, Training officer of the Mountain Rescue Committee of Scotland, founder of Climb Mountains UK, friend and confidant; a mountaineering expert.
- Dave Whalley MBE BEM, RAF Mountain Rescue Team Leader with over 30 years experience and who has an unsurpassed knowledge and experience of SAR to which I can only ever aspire.
- Sigurður Ólafur Sigurðsson, Director of ICE-SAR Landsbjörg Rescue School Reykjavik, Iceland where professionalism is seamlessly integrated into all aspects of SAR and who arranged the excellent glacier trips with various Icelandic SAR teams and also for providing the photographs used in the **GNSS in Emergency Management** section (p. 000–00).
- Lynne J. Engelbert, Section Chief of Training, NASA, who worked hard to allow me to visit NASA and introduced me to the Disaster Assistance and Rescue Teams.
- Carole Smith, Directeur of the National Search and Rescue Secretariat, Government of Canada, who gave me the opportunity to present at SARSCENE and work with Canadian mountain rescue teams (MRT) and coastguard.

- Hiroshi Sato, Leader of Special Rescue Teams, Tokyo, Japan, for the tremendous experience of working with both MRT and coastguard teams on live missions and accompanying me to Mount Fuji.
- Hewitt Schlereth is a retired NAMS certified marine surveyor, long-time sailor, and authority on celestial navigation with six books published on this subject. His expert guidance and support have been invaluable as has his sense of humour.
- Majid Sabetzadeh, international SAR instructor, Specialist in Middle Eastern SAR, Team Member of San Diego SAR.
- Keith Lober, Emergency Services Director of Yosemite SAR, USA. A totally dedicated team leader, eager to share best practice and enthusiastic to take on new ideas and concepts working in the wilderness. Also a special thanks to John Dill of YOSAR.
- Scott Amos, Jen Changleng and Gavin Kellet of Tweed Valley Mountain Rescue, fellow team members, great friends and all superb navigators who helped keep me focused and on track!
- Jim Gilbreath, Navigation Instructor of the Mono county SAR Team, who shared his phenomenal navigational experience with me in the Switzerland of America, Rock Creek in the eastern Sierra and whose expert eye for detail helped me proof the book.
- Geoff Summers MBE, one of the world's most accomplished polar travellers who has crossed Antarctica by its greatest axis and has been the Safety Officer on a Russian nuclear-powered ice-breaker. He kept me ontrack in a very difficult section.
- S.O. Vladimirov, Chief of the Department of Space Navigational Systems, Connection and Ground Complexes of Navigation, Roscosmos, Russia, for unique access to the Russian Global Navigation Satellite System, GLONASS.
- Sir Ranulph Fiennes BT OBE, British adventurer and the first man to visit both the North and South poles by surface means and to completely cross Antarctica on foot – simply an inspirational human being.
- Sean Kirsch, Staff Sergeant, 131 Pararescue Squadron, United States Air Force Special Operations Command, Moffet Federal Airfield, California, USA. A kindred spirit at the cutting edge of navigating in all terrains and expert responder.

Also my thanks to the many manufacturers who provided the equipment to test and gave me technical support when I needed it: Anquet, Garmin, Keela, Steiner-Optik and Suunto.

WHAT IS MICRONAVIGATION?

Most people will be quite confident of their navigation skills on a warm, clear, summer's day or while following a well-marked footpath. The problem is that these scenarios often do not last long: the weather can change in an alarmingly brief time, signs run out, tracks disappear or are impassable.

There is only one way to be really confident in navigating in all conditions, and that is to learn the techniques of micronavigation which are similar to the techniques used in orienteering. With practice, you will soon be able to navigate to all sorts of tiny features on the map – even the smallest kink in a contour line – in poor visibility from fog to night-time. Micronavigation is a key skill for anyone who wishes to navigate competently and safely.

Micronavigation is an easy-to-learn technique of land navigation. You navigate in a series of small distances, called legs, and focus on the immediate features in your landscape to continually ascertain position: this minimises the chance for error.

All navigators make small errors in compass reading, correcting for magnetic declination, and then in both judging distance travelled and how far they have drifted off course. The further you travel on a bearing the greater the error. So by employing short legs these mistakes are continually corrected.

Micronavigation builds upon your innate ability to create a greater awareness of your immediate environment, using your senses of sight, sound, smell and touch and relating this information to the use of robust and reliable non-specialist equipment: a map and baseplate compass.

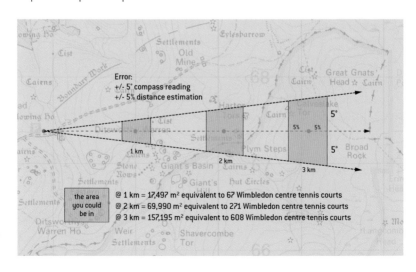

Error:
+/- 5° compass reading
+/- 5% distance estimation

5°
5% 5%
5°

1 km
2 km
3 km

the area you could be in	@ 1 km = 17,497 m² equivalent to 67 Wimbledon centre tennis courts
	@ 2 km = 69,990 m² equivalent to 271 Wimbledon centre tennis courts
	@ 3 km = 157,195 m² equivalent to 608 Wimbledon centre tennis courts

Micronavigation can be used in the most difficult of terrains including **Alpine, Arctic, Desert, Forest, Jungle, Mountain, Shoreline and Urban** and is employed by specialists varying from search and rescue (SAR) teams to the Special Forces.

Micronavigation can also be augmented with, yet is not dependent upon: altimeters, celestial navigation, environmental navigation and global navigation systems (GNSS).

Different types of navigation may be used in isolation, but just as the Phoenicians relied heavily upon celestial navigation augmented with environmental navigation, today the expert land navigator utilises and combines as many different types as possible.

↘ EXPERT TIP

→ The more navigational techniques you carry with you in your mind's rucksack the more capable and adept navigator you will be.

Using a dried-up riverbed as a linear feature.

HOW TO USE THIS FIELD INSTRUCTION MANUAL

Follow the instructions in this manual to the letter and you can become an outstanding navigator on land. **Do not skip or jump sections**: they have been developed from teaching basic navigation to groups of beginners, to advanced navigation to experienced SAR professionals. They work!

This manual takes nothing for granted and you need no prior knowledge of navigation whatsoever.

In the first chapters of this book you are going to build 'learning scaffolding' and construct a simple framework upon which you can further develop your knowledge and skill sets with the more challenging techniques and theory.

- **Words styled in bold grey type have full explanations elsewhere in this manual.**

- *Sentences in bold italic are essential learning – learn them by heart.*

- *EXPERT TIPS* are the handy hints and snippets of information to help you to be spot on in your navigation.

- *EXPERT FACTS* are the pieces of information that will help you understand and remember the theory sections.

- *QUIRKY FACTS* are fun but also help you remember the section.

There are only three home-study topics. Theory is kept to an absolute minimum and then only that which is absolutely necessary. Take time and learn these topics: **The Three Norths, Mapping Systems** and **GNSS/Satnav Introduction**.

Basic Techniques are highlighted to give you the basics you require to navigate. The more you learn beyond this the greater your repertoire of navigational skills and the more proficient and safer navigator you become.

There are many different types of maps and numerous navigational pieces of equipment from compasses to satnav. This manual is not a discussion document and recommends only map types and equipment which has been used extensively by the author throughout the world, both for pleasure and in SAR.

It dispenses with unnecessary theory and technical jargon. You will only learn what you need to know to **micronavigate** competently and safely.

Sometimes written instructions will be short and instead techniques will be described using step-by-step photography by experienced navigators. This should stimulate your understanding with visual clues, in addition to the written word – this can be a very powerful method of learning.

There are seven specialist terrain sections. These are environments where additional techniques and sometimes additional equipment may be needed to navigate safely. In each specialist terrain section, techniques are explained fully to obviate the need for unnecessary cross-referencing and backtracking for the user.

Easy to use: easy to learn – helpful hints are included on what to do if you get stuck learning a technique and how to make the technique straightforward to use in the field.

↘ EXPERT TIPS

→ Tell someone where you are going and what time you are going to be back and at what time they should call the emergency services if they cannot get hold of you.

→ Check the weather forecast and wear the appropriate clothes and footwear.

→ When you have mastered your new skills, you should experiment with different types of maps, using a variety of scales because your skills will be easily transferable.

→ The background, theory and history of navigation and its development make interesting reading – my favourite book on this subject is Dava Sobel's *Longitude* (Fourth Estate, 1998).

→ Navigational terms change not only from country to country, but also from group to group.

→ Check the **micronavigation.com** website forum for updates and discussion on new and existing techniques.

↘ WARNING

Do not proceed straight into an unknown area and attempt to learn these techniques: instead, read through the techniques you are going to learn and practise them in a safe environment that is well known to you – such as gardens, parks or local walks.

Using a mountain ridge as a linear feature.

CONVENTIONS/ ABBREVIATIONS

Conventions are the terms and symbols used in the manual and you need to know and understand them.

Key navigational terms

- **Route**: from the start of your journey you will have a destination you wish to reach, this is called the **Objective**.
- To reach your objective you navigate a series of short **Legs**.
- Each leg starts from a known point and leads to an identifiable point on the map – this is called an **Attack Point**.
- **Technique** is a single method used to navigate.
- **Procedure** is a combination of techniques used to navigate.
- **Feature** is an object or landmark such as a building, cairn, summit, river, forest or lake.

X = *Objective*    = *Legs*   ◯ = *Attack Point*

Measurement, units and abbreviations

The metric measurement system is primarily used in this manual.

Metric
Kilometre – km
Metre – m
Centimetre – cm
Kilometres per hour – kph
1 Kilometre =1,000 m
1$^1/_2$ Kilometres – abbreviated as 1.5 km
$^1/_2$ a Kilometre – abbreviated as 500 m
Celsius – °C

Imperial
The following abbreviations are used:
Feet – ft
Yards – yds
Miles – miles
Miles per hour – mph
Fahrenheit – °F

Positions
⊙ **Fixed Position** is when you are confident of your position and can pinpoint it on a map. Either drawn as ⊙, or written down as 'Fix'.
△ **Estimated Position** is where you know the general area you are in but cannot pinpoint your exact position on a map, Either drawn as △, or the initials 'EP'.
+ **Dead Reckoning** is a position which has been calculated from your start or last attack. The further you travel the less accurate it becomes. Either drawn as +, or noted on the map as 'DR'.

Azimuth
The azimuth is taken to mean the horizontal angle of a **Bearing** clockwise from north.

- The graduations on your compass bezel in degrees read from 0° to 360°.
- Degrees can be subdivided into 60 minutes and minutes can be subdivided into 60 seconds. Seconds are too small a unit to be required in general navigation.

1° (degree) = 60' (minutes)
1' (minute) = 60" (seconds)

The only time you will encounter minutes using your compass is when referring to the **Magnetic Declination** stated on maps, which for practical purposes, you will round up to the nearest degree:

- 005°31' will be expressed as 006° as 31' is more than half of 60 – which is the number of seconds in a degree.

- 005°29' will be expressed as 005° as 29' is less than half of 60 – which is the number of seconds in a degree.

Minutes and seconds of azimuth are also used in the coordinate system of latitude and longitude where they are usually written as X°X'X" — see **Mapping Systems**

Global Navigation Satellite Systems (GNSS)
Handheld navigational equipment which uses satellites (including those which use the ubiquitous, American GNSS – GPS) will all be referred to as **satnav**. The two types of satnav receivers used are **mapping satnav** and **basic satnav**.

Obtain as clear a view of the sky as possible when using satnav.

EQUIPMENT LIST

I have tested dozens of pieces of equipment throughout the world, often operating in the most extreme conditions. The items listed here are the ones that I have trusted with my life. I believe they are the best of their kind, being reliable, robust and efficient.

Essential equipment

1 Suunto M3 Global compass – magnetic compasses are the mainstay of land navigation because they are relatively low-cost, durable, and require no maintenance or no electrical power supply. The M3 is a dependable, high-quality baseplate compass which performs flawlessly in even the most demanding of conditions from –40° C to +60° C. The notched bezel is easy to turn, even in cold conditions while wearing gloves, it has easy-to-read luminous markings and a magnifying lens, making it ideally suited to navigating in conditions of poor visibility as well.

2 An up-to-date, good-quality topographic map. National agencies such as USGS in the USA and Ordnance Survey (OS) in Great Britain produce such quality maps. If you can, buy laminated or plastic maps; these are ideal since they are waterproof.

3 Grid Reference Tool – an inexpensive means of easily taking accurate grid references in the field. (Co-developed by the author for use in SAR and general navigation.)

4 Tally counter – measuring distance while moving is an indispensable navigational technique and accuracy is vital. In one technique, called **Pacing**, you need to count your paces – counting them in your head distracts you. This little tool will do it for you.

5 Grease pencils/chinagraphs are great for writing on waterproof maps, as the information can be rubbed out when you no longer need it.

6 Head torch – you should never navigate at night, even when natural light is good, without a head torch; weather conditions are liable to change at any time and clouds can obscure light from the moon and stars. Chose a headlamp with an LED bulb(s): these use less power, last longer and are more durable than conventional bulbs. Select one with a focusable beam (a minimum range of 50 m on full beam) for spotting features in the distance, and which has different setting levels so when reading the map the light can be set to low to reduce the interruption to your natural night vision.

7 Backup set of lithium disposable batteries (see **Battery Selection** for a comprehensive overview of battery types and recommendations).

↘ EXPERT TIPS

→ Attach a length of paracord to every item with a loop at the other end that the item will fit through so that you can easily fasten/unfasten it to your rucksack/jacket for security. Seal the cut paracord over an open flame.

→ Always take water, a mobile phone with fully charged batteries and a whistle.

→ Try to choose electrical items which all use the same battery type/size so you don't have to carry lots of spares!

→ Take spare reading glasses, if you use them. If you forget them, use the magnifying lens on your compass.

→ Choose a personal locator beacon (PLB) with GNSS that can transmit accurate latitude and longitude coordinates on activation.

Adding the cost of these seven items equates to the cost of a meal out for two, which is relatively inexpensive – considering that your life may depend upon them!

Desirable equipment

- GNSS – accurate and reliable handheld satnavs are relatively inexpensive. (Reasonably new second-hand units are a good consideration and about the same price as your compass.) Depending upon your specific needs, there is a range of handheld satnavs. All of them give your **Location**, can create **Waypoints** and follow **Routes**, record **Tracks** and have a 'Trackback' facility. See **GNSS** section.
- A backup set of batteries for your satnav (lithium, see **Battery** section).
- If your maps are paper, then a waterproof map holder/cover is advisable.

- Waterproof paper, especially if you are printing out your own maps/GNSS route and stellar navigation tables.
- Waterproof notebook – write using a pencil in all wet or dirty conditions.
- 55 m of paracord – it is quick drying, rot and mildew resistant and relatively inexpensive. You use it to secure your compass and satnav (rather than the short, flimsy ones supplied by most manufacturers) and also to measure distance in techniques such as **The Wheel**.
- A backup compass and satnav. An old baseplate compass and a second-hand basic satnav are fine so long as they both function correctly.
- Pacing and timing card.
- Slope angle card.
- Walking or ski poles. If you use these, you can easily mark them to estimate slope angle by marking up one pole into eight equal lengths.

7° slope

37° slope

❶ Put the two poles side-by-side and ensure they are the same length.

❷ Measure the pole and divide into eight equal lengths using insulating tape to mark each one.

❸ From first mark down from the top of the pole write in indelible pen:

41° → 37° → 32° → 27° → 20° → 14° → 7° (The top of the pole is 45°).

You have now created a tool for measuring slope angle that can be used to determine:
A. snow slopes prone to avalanche – generally avalanches trigger on slopes ≥30° and slopes of around 40° are the most dangerous;
B. your elevation (height above sea level).

Equipment for SAR and some special environments

- Ruggedized mobile phone.
- Small 406MHz PLB (personal locator beacon).
- Quality pair of compass binoculars.

PSYCHOLOGY

Clear thinking is essential for accurate navigation. Tiredness affects your judgement and panic leads to rushed and bad decisions.

Prevention is the answer so:

1 Take plenty of fuel, in the form of food, and fluids.

2 Keep warm by wearing and packing gloves, hats and so forth.

3 Do not wait until the mist descends; keep prepared by building up a mental picture of your route and constantly compare it with the map.

4 Double-check everything by either confirming what others have told you or estimating a bearing before you take it.

5 Trust in your map and compass.

Deep inside the Arctic Circle,
using the sun's position to orient
the map.

SECTION ONE
THE ESSENTIALS

ENVIRONMENTAL NAVIGATION

From infancy we learn to navigate our environment using all of our senses. Most of the techniques taught in this manual simply mirror innate skills.

Introduction

You will already have a wealth of navigational tools at your disposal which you have been developing and adding to since birth. For example, collecting features that confirm your route is a technique you probably employ everyday travelling to and from work — from noting significant traffic junctions, to registering the noise of the train brake as you approach the station from where you will continue the next leg of your journey. My Day One objective is to break with convention and reawaken these innate skills in the natural environment.

> → A major step to becoming a skilled navigator is when you stop thinking
> about yourself as the reference; instead relate the world around you and the
> orientation and position of everything in it.

To instruct navigation is simply to formalise these skills and extend them with the mastery of man-made tools. One of the single biggest mistakes that people can make when they teach formal navigation with a **Map** and **Compass** is to focus on the use of these tools and to ignore the individual's natural skills.

Learn to use the wealth of information your surroundings offer you in helping determine where you are. Spend time re-familiarising yourself with the natural environment before you start working with the rest of this manual; a few extra days will make little difference to your overall timetable and learning to read environmental clues is a key component of micronavigation – they are invaluable.

To start with, take a walk on a route you are very familiar with and run through the following check list in the field until it becomes second nature. You will be surprised what else you become aware of – add your own observations.

Wind direction

Wind direction often changes frequently over the period of a day so look for permanent signs of the prevailing wind's direction, especially trees and shrubs whose branches are bent to shape by the directional wind – they can act as **Radial Arms** and help you determine the other cardinals of the compass.

Prior to entering an area, determine which way the prevailing winds blow and if they are seasonal. This can be found easily from many sources ranging from meteorological internet sites to local farmers.

Wind speed

Wind speed generally increases the higher you climb and a quick technique to predict the speed in the UK is to count the isobars, if they are 4mb apart (as in Met Office forecasts), covering the UK and allow 16 kph (10 mph) for each isobar: this will be your rough wind speed at about 300 m (1,000 ft) above sea level.

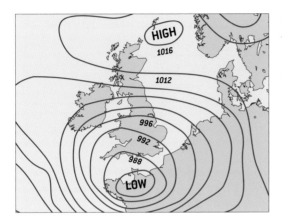

In this example (seven isobars fall across the length of the UK) you should expect winds up to 70 mph (112 kph) at 300 m (1,000 ft).

Scent

Face the direction the wind is blowing, close your eyes and smell the air. Think carefully about the scents you can detect:

- traffic fumes suggest a road
- the scent of a forest
- farmyard smells
- industrial/factory fumes
- in urban environments, a tube station/subway entrance.

Now look at your map in the direction of the blowing wind and see if any of these features are present.

Temperature

As you gain height the temperature of the air decreases. The amount it decreases depends primarily upon the amount of water vapour in the air and the extremes of these are:

- dry air found in deserts and high mountain ranges, temperature changes at an average rate 10° C/1,000 m (<10% relative humidity)
- clammy, muggy air found in rainforests, temperature changes by an average of 5° C/1,000 m (saturated air with >90% relative humidity).

Think about where the climate in the region you will be navigating in fits between the two extremes given above and remember that as you lose height the temperature increases by the same values.

In winter conditions look for where the frost has melted on rocks, this will be the side where the sun has shone the most.

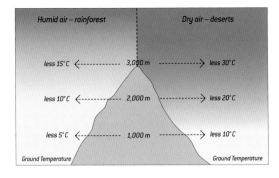

Humid air – rainforest		Dry air – deserts
less 15°C ←--------	3,000 m --------→	less 30°C
less 10°C ←--------	2,000 m --------→	less 20°C
less 5°C ←--------	1,000 m --------→	less 10°C
Ground Temperature		Ground Temperature

The rate-of-change of air temperature with altitude is primarily governed by the relative humidity of the air and ranges between the two extremes of rainforests and arid deserts.

Vegetation

Take note of what you have walked over – a marsh, sand, grass or heath – and compare it to the vegetation marked on your map. Look for the edges of forests and forest clearings, also the type (coniferous or deciduous), and relate these to where you think you are on the map. Learn the symbols for these different types of vegetation (**refer to the legend at the side of your map**).

Different plants grow in different habitats and at different altitudes. Prior to embarking upon your trip learn about the plants of the region you will be visiting, how to identify them and where they are likely to be found. At the highest elevations, trees cannot grow and vegetation becomes alpine. Just below the tree line, one may find subalpine forests of needleleaf trees, which can withstand cold, dry conditions. In regions with dry climates, the tendency of mountains to have higher precipitation as well as lower temperatures also provides for varying conditions.

Remember, though, that landscapes change with time, so check your map's age, and consider things that may have altered the environment, such as the clearing of land for new developments, forest fires, farming usage and so forth.

Geology

Before your journey, mark areas of different geology on your map – this will help give you an indication of where you are? This information is readily available on the internet.

Knowing the geology of an area can help you understand what is under your feet and where you might be.

☐	Granite	☐	Slates and shales - Carboniferous
☐	Cherts, N - Meldon, E - Teign	☐	Tertiary sands and clays
☐	Dolerites	☐	Slates and shales - Devonian
☐	Limestones, N - Carboniferous E - Devonian	—	Fault lines
☐	Volcanics	- -	Metamorphic Auredele Outline

Sound

Can you hear streams, rivers, waterfalls or traffic? To enhance your hearing cup your hands behind your ears, it will help you identify the direction of the sound. This effect can be dramatic, detecting sounds which ordinarily you cannot hear. With practice this method of direction-finding can be accurate to within ± 10°.

- Cup both hands behind your ears.
- Rotate your body, moving your feet and keeping your head fixed forwards, towards the sound.
- Make small adjustments by moving with your feet to where the sound is most intense.
- Keep standing and take a bearing in the direction you are facing.

While navigating in dense woodland, searching for a lost party who were giving six whistle blasts repeated at five-minute intervals, this method was used to find them.

Aircraft

You can either identify commercial aircraft flying overhead using your binoculars or,

more commonly, by the contrails (short for 'condensation trails' sometimes called vapour trails or distrails) they leave. The exhaust from aircraft engines create these artificial clouds of condensed water vapour and, depending on atmospheric conditions, they remain observable for anything from a few seconds or minutes, to many hours. All commercial aircraft fly along predetermined air corridors on specific routes and frequently these run in one direction only. As a result, if you know the direction of these flight paths you can use the contrails to determine the other cardinals of the compass. Prior to embarking upon your journey, check which flight paths exist for aircraft in the area. You can find these on internet sites such as **www.flightradar24.com**. Military aircraft do not use these corridors and tend to fly lower than commercial aircraft.

Aircraft lighting

All aircraft are required to have a green navigation light on the right side, a red navigation light on the left side, and a white navigation light on the tail which must be on from sunset to sunrise. Additionally, most commercial aircraft have bright flickering white strobe lights (to light up clouds) which can be seen from the ground. Using these lights you can determine which direction that aircraft is moving at night and, if you know the air corridors, you can determine the other cardinals of the compass.

Aircraft obstruction lights

The proliferation of wireless communications has resulted in the building of many broadcast masts and towers. These are clearly visible both during daylight hours and at night as they are usually situated on the highest elevation of land in the area. Consequently they are excellent to use as **Attack Points**, **Radial Arms** and when calculating distance from/to.

International Civil Aviation Organisation rules stipulate that all structures above 60 m (200 ft) must use night-time aircraft obstruction lights. Typically these are red (constant or slow flashing) and situated on the top of these structures and at predetermined intervals down it.

As most of these towers are a relatively recent introduction, with new ones continually being erected, this is another important reason to use up-to-date maps. Where a structure is not marked on your map take the time to add their location and their height (using your grease pencil or on your digital mapping). Government sites such as **www.sitefinder.ofcom.org.uk** detail both their exact location and height.

Other tall structures usually located on high elevations include water tanks and wind turbines. Aircraft obstruction lighting is also used on electricity pylons, chimneys, tall buildings and cranes.

Pylons

Many overhead power lines run for great distances in the same direction and can be used as **Parallel Features**, to **Handrail** beside and for **Radial Arms**, prior to embarking upon navigating study a map of the area and search for these (see **Techniques** section). In addition, there are common types of pylon which are all exactly the same height. By learning to recognise these you can use them to judge distance to/from.

Sun, moon and the stars

A whole section of this book is dedicated to the numerous techniques involving the sun, moon and the stars – **Celestial Navigation**.

Bird and animal movement

Check the times of year for migrating flocks of birds – there are numerous websites where you can find this information. Learn to identify them and know which direction they are likely to be flying in.

If there are migrating animals in the region you will be navigating, again find out which species you are likely to see and the routes they follow.

✘ **QUIRKY FACT:** Researchers at the University of Duisburg-Essen in Germany have found that herds of cattle and deer tend to align themselves $\pm 5°$ from a line from north to south.

302779.

→ EXPERT TIPS

- → A westerly wind blows east, northerly blows south and so forth. It is better to stand and feel which way the wind is blowing, turn to face it with your map set and see what is ahead of you on the map.

- → Professional trackers continually hone and refine their ability to read environmental clues.

- → Buy books on the local plants and trees of your area; they are interesting to read and help bring your environment alive.

- → A bright white light seen travelling across the sky at night may be the Space Station! See **www.heavens-above.com**.

Clouds

The surface of the earth's reflection on the bottom of clouds can indicate the type of terrain beneath them.

If the natural light is failing, search the sky above the horizon for a patch of light which could be the reflected light from a small town – the light pollution from a small town at night is visible from 15 km, the light from a city from 50 km.

In the arctic regions the colour of the underside of clouds gives you clues as to the type of land beneath:

- **Black** – open water, snow-free ground and forestation
- **White** – snow fields and sea ice
- **Grey** – new ice
- **Mottled grey** – pack ice or drifted snow.

Places of worship

Religious buildings are usually aligned in a specific way to the cardinals of the compass. Christian churches are generally aligned east–west, with the high altar on the eastern side. Most mosques contain a niche in a wall that indicates the *qiblah*, which is the direction of a point in Mecca towards which a Muslim prays during *salah*. The latitude/longitude of this point is 21°43'N 39°06'E – so before you venture out, check on Google Earth where it is in relation to the area you will be navigating in.

MAPS

You are going to learn to work with topographic maps, all of which share the same basic features.

Introduction

Maps are the most essential tool used to navigate. Much of the time navigation can simply be a matter of using a map alone, by relating what you see on the map to your surroundings.

Whether printed or digitally displayed, a map is a representation of features and objects from rivers and mountains, to communication towers and churches. They are all drawn to scale, so the distances between features on the map are in proportion with the actual landscape.

These basics are intended to be applied to topographic maps from British OS, American USGS and German Bundesamt für Kartographie und Geodäsie, to Swiss Swisstopo, Australian NATMAP and military maps … in fact any topographic map!

Map types

Maps come in four formats:

- Printed maps
- Digital maps
- Satellite and aerial photographs
- Custom maps.

Printed maps

Every country has its own particular mapping and grid system that are based upon projecting their region of a spherical world onto a flat surface; because of this, all maps will have small inaccuracies.

A printed map is a 'birds eye' (looking down), two-dimensional (flat) representation of a curved surface (as the world is a sphere).

Topographic maps have lines of elevation (height) drawn on them which can, with practice, be easily interpreted by the reader to create a three-dimensional image of the land. (**Contours** see p. 50–1)

They also show both natural and man-made features such as roads, rivers, buildings, often the nature of the vegetation, and the names of mountains, hills and districts.

Maps are based on their own grid north, which is usually different to either true north or magnetic north, and when using a compass you need to make adjustment for this difference (**The Three Norths,** see p. 64).

Most countries have their own national mapping agencies, such as Ordnance Survey (OS) in the UK and the United States Geological Service (USGS) in the USA.

In addition, specialist cartographers create maps for walkers, hikers and mountaineers such as Harvey Maps in the UK (**www.harveymaps.co.uk**). These maps are compiled from Harvey's own original aerial surveys (not redrawn from OS maps) and then field-checked by experienced surveyors. The presentation emphasises contours and topographical features while the subtle use of colours makes these maps very easy to read in the field.

> ✕ **EXPERT FACT:** We need to think of digital maps as **Location Data** which we can customise.

Digital maps

Digital maps are dealt with in detail in the **Digital Mapping** and **GNSS** sections. In summary they look the same as printed maps but are displayed as an image on an electronic device. In practice there is a whole lot more you can do with them.

National mapping agencies, such as Ordnance Survey and United States Geological Survey, allow access to their maps online: **leisure.ordnancesurvey.co.uk/products/ digital-maps** and **nationalmap.gov/digital_map** respectively.

→ EXPERT TIP

→ It is worth remembering that while many mapping sites contain national data, they do so by keeping licensed copies of the data on their own servers, rather than direct links to the national mapping agencies website. This means that this data can be perhaps a year or two out of date on these third-party websites. For the latest definitive mapping data, always go to the national mapping agencies free map website.

→ THE MAPPING REVOLUTION

It seems strange to talk about an enormous transformation in something as well-established as cartography, yet this is exactly what is happening. The revolution is the shift from paper maps to paperless maps – **Digital Mapping** in other words.

Conventional map data is compiled and formatted into a virtual image, which can then be used in computer applications such as Google Earth and also loaded into electronic navigation devices such as car satnavs and handheld satnavs.

Digital mapping also allows for a transformation from simple greyscale contour maps to full-colour 3D maps which can be virtually toured.

In recent years, digital mapping has come into its own allowing the user to customise their maps. Just a few examples are:

- Car satnavs, where locations such as home or work are marked.
- South Wales Fire Service, UK, have added features critical to their work, such as fire hydrants and detailed building plans to this database.
- In my Mountain Rescue (MR) team we have added all gates and tracks through fields and across the mountains where a stretcher can be carried – over 2,000 of them.

Furthermore, digital map locations can include popup/dropdown windows with additional information about the feature. So, for example, the Fire Service's building plans have information about hazardous substances that may be present, the identity of any key-holders – and because they are paperless they can be continually updated.

Many smartphones come with inbuilt satnav. You can view on their screens a map that can show where the nearest public defibrillator is in relation to where you are, the nearest bus stop, ATM, pharmacy, cinema. The list of bespoke information which can be added to virtual maps is almost endless.

There are free editable maps of the whole world available online, see **micronavigation.com**, which are created by anyone who wants to contribute, in the same way as Wikipedia. These sources allow you to view, edit and use geographical data in a collaborative way from anywhere on earth.

The car satnav manufacturer TomTom have an online community which receives thousands of corrections every day from its millions of users, not just updating road changes but also intelligent routing based on real traffic speeds.

Satellite and aerial photographs

With the advent of Google Earth and Microsoft Virtual World, aerial maps are available to anyone with access to the internet and they now constitute a significant component of navigation, especially when used in association with **GNSS**.

- A recent aerial photograph shows changes that may have taken place since the topographic map was made.
- In **Forest** areas where either felling or new planting has taken place.
- In **Urban** areas, principally on the fringes of cities, towns and villages, where new

housing may have been built and also around industrial sites showing new units.

- All regions where new roads are either under construction or have been built.
- Many military features are not shown on topographic maps.
- Disaster areas for SAR, from flooding to plane crash sites, which can be available within hours of the incident and continually updated.
- Remote areas of the world, such as **Jungles**, which may not as yet have been mapped.

The drawbacks to aerial maps are:

- It is not always easy to ascertain how old internet images are.
- The scale and position location are estimates.
- Features can be difficult to identify without map symbols.
- Contours are difficult to interpret without overlapping photographs used in combination with stereoscopic viewing equipment.
- Colour contrast and tone can be very similar, unlike a topographic map, making them difficult to use in poor visibility.
- To interpret aerial maps expertly takes a great deal of training compared to topographic maps.

When **Aerial Photographs** are used in conjunction with **Topographic Maps** they provide the most accurate detail of the terrain for navigation possible.

Custom maps

This is a revolutionary and exciting development, made possible by the advent of the internet and the development of digital mapping. These days you are able to take any map, whether paper or a digital image downloaded from the internet and use it to navigate with on your satnav mapping receiver. See **GNSS Advanced** section.

In summary, digital maps, satellite and aerial photography, and custom maps are these days all potentially important navigational aids — yet they cannot replace conventional printed topographical maps; instead they add significantly to their effective use.

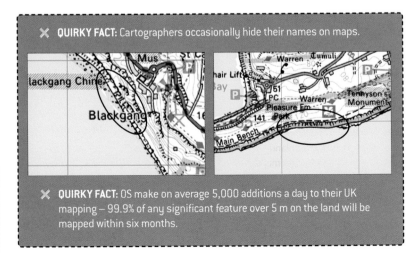

✕ **QUIRKY FACT:** Cartographers occasionally hide their names on maps.

✕ **QUIRKY FACT:** OS make on average 5,000 additions a day to their UK mapping — 99.9% of any significant feature over 5 m on the land will be mapped within six months.

Scale

Let's get the tricky part out of the way now: map scales! Scale is the relationship between distance on the map and distance on the ground. It is given as a fraction or a ratio – 1/25000 or 1:25 000 – and is always printed on the map.

- **Example:** A map with a scale of 1:25 000 means that every one unit of measurement on the map (like a centimetre) is the same as 25,000 of those units (in this case 25,000 cm or 250 m) in real life. If the scale is imperial rather than metric it makes no difference: on a 1:63 360, 1 inch on the map would represent 63,360 inches (1 mile).

The first number (map distance) is always 1. The second number (ground distance) is different for each scale – the larger the second number is, the smaller the scale of the map. 'The larger the number, the smaller the scale' sounds confusing, but in fact it is easy to understand.

In the small-scale map (such as 1:250 000) there is less room; therefore, everything must be drawn smaller, and some small streams, roads, and landmarks must be left out altogether. On the other hand, the large-scale map (1:25 000) permits more detail but covers much less ground.

- *Large-scale maps show small features on the land, such as an individual house.*
- *Small-scale maps show large features on the land, such as an entire town.*

Maps are published at different scales and it's important to choose the right scale for the task. In the UK, OS produce outstanding maps. Their 1:25 000 scale maps are called Explorer maps and there are 403 printed versions which cover the UK. Their 1:50 000 maps are called Landranger and 207 of these printed maps cover the UK.

Explorer maps show minor paths, field boundaries (walls and fences), open access areas and public rights of way (except in Scotland), and small areas of marshland, rocky ground and small streams: Landranger maps do not show these.

However, Landranger maps do have their place in walking and mountaineering; indeed, some other Scottish MRT use these as standard issue where fences and rights of way are unimportant and where they need to view larger areas of land. A few

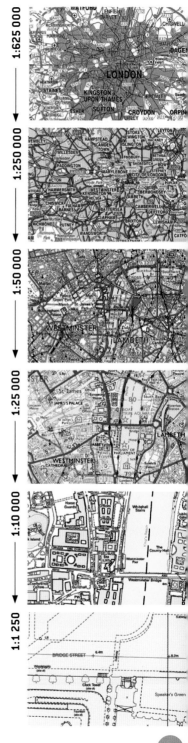

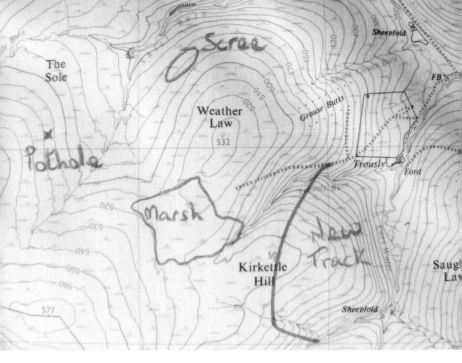

years ago I trekked across the Cairngorm Plateau and I used Landranger maps quite simply because it meant I had less to carry!

My MRT use Explorer (1:25 000) maps as standard issue. A number of the northern Scottish MRTs I work with prefer the Landranger maps, and Icelandic teams use 1:100,000 scale when working on the vast glaciers and icecaps there. It is really down to personal choice, how complex the terrain you are working in is and how large an area you need to search. I personally prefer to use 1:25 000 scale maps and they are mainly used in this book. In towns 1:10 000 or even larger-scale maps are more appropriate.

When I am instructing teams, I place a great deal of emphasis on manually personalising and updating maps; you should do so too. Annotate your maps, indicating streams that have dried up, paths that are incorrect, a rock fall or wash-out, unmarked potholes, new tracks, overgrowth, scree and so forth.

→ EXPERT TIPS

→ Buy the most up-to-date maps you can: from the minute they are produced maps start going out of date.

→ 'Tourist' features in blue ink (Nature Trails, Visitor Centres and – importantly – ski-lifts) are not placed accurately. They should **NOT** be used as navigation aids.

→ When navigating across national boundaries, be careful if using maps from different countries. For example, the French tend to use 10 m contour intervals, whereas the Swiss use 20 m intervals – so what looks like a navigable slope on a Swiss map, if you have been using 10 m intervals, is in reality twice as steep.

MAPPING SYSTEMS

Maps are a graphic representation of the real world. Cartographers project images of the earth onto flat surfaces – these are called mapping systems.

First of all let's clear up a common misunderstanding about the difference between a coordinate system and a grid system:

- a coordinate system describes an exact location – **RED CIRCLE**
- a grid reference places you inside a grid square – **BLUE CIRCLE**.

The coordinates for the red circle are 3,3 – it is an absolute position. The grid reference for the blue circle is 1,4 – it would place you anywhere inside this small square. For a detailed explanation of this see **Using Grid References and Coordinates.** We use coordinate and grid reference systems to describe a point on the earth's surface.

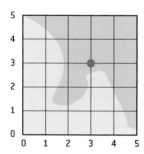 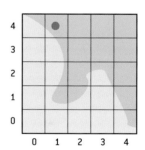

Global mapping systems

There are three main systems used for the world and specifically one for the poles.

- **Longitude and Latitude**. Extensively used in aviation and maritime navigation, they are also used in land navigation.
- The **Universal Transverse Mercator** (UTM) covers all of the earth with the exception of the Arctic and Antarctic.

✖ **WORLD GEODETIC SYSTEM (WGS84):** A map datum is a three-dimensional model of the earth with the centre as its origin on which mapping, grid and coordinate systems are based. The earth is not a perfect sphere; instead it is shaped more like a beach ball that is being slightly squeezed (at the North and South poles, making it fatter around the equator) – a shape that is known as an ellipsoid. Map datum have been created to represent this shape and include a nominal average sea level, but they do not model mountains and valleys. The WGS84 is the underlying datum for most maps, every satnav and is the world standard for digital mapping.

- **Universal Polar Stereographic** (UPS) uses the same conventions as the UTM and covers the Arctic and Antarctic only.
- **The Military Grid Reference System** (MGRS) is derived from the UTM and UPS grid systems, but uses a different labelling convention.

All these coordinate and grid systems have the North Pole at the top.

Latitude and longitude

This is the universal coordinate system used the world over. All air traffic and maritime vessels use it, plus many of the emergency services.

Latitude

- All lines of latitude run horizontally around the globe
- lines of latitude run parallel east to west
- used to express how far north or south you are, relative to the equator
- shows your location in a north–south direction
- latitude is an angular measurement in degrees from 0° at the equator (low latitudes) to 90° at the poles (+90° N for the North Pole or -90° S for the South Pole)
- lines of latitude are always the same distance apart – 1° = approx. 110 km
- abbreviated to *Lat*
- sometimes called parallels.

Longitude

- all lines of longitude run vertically on the globe and converge at the poles
- start at **True North**
- shows your location in an east–west direction
- longitude is given as an angular measurement ranging from 0° at the Prime Meridian (also known as the International Meridian at Greenwich, England) to +180°E and –180°W
- abbreviated to *Long*
- sometimes called meridians.

Long/lat is measured in degrees (°), minutes (') and seconds (").
60 seconds = 1 minute; 60 minutes = 1 degree.

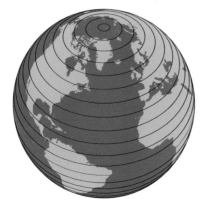

Latitude – equator in red

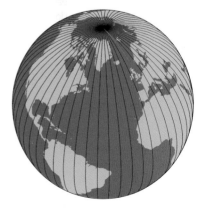

Longitude – International Meridian in red

A specific longitude may then be combined with a specific latitude to give a precise position on the earth's surface which is known as 'absolute location'.

Latitude and longitude is generally expressed in three ways. The latitude is always given first followed by the longitude:

Venue	Latitude	Longitude	Used by	Absolute location written as:
WHO HQ Geneva	46°13'58"N	6°08'04"E	Aviation/ Marine	46°13'58"N 6°08'04" E
	46°13'57.78"N	6°08'03.79"E	Google Earth	46°13'57.78"N 6°08'03.79"E
	+46.232777°	+6.134444°	MSN Virtual Earth	+46.232777° +6.134444°
UN HQ Manhattan	40°45'05"N	73°58'04"W	Aviation/ Marine	40°45'05"N 73°58'04"W
	40°45'04.59"N	73°58'03.89"W	Google Earth	40°45'04.59"N 73°58'03.89"W
	+40.7513888°	−73.967777°	MSN Virtual Earth	+40.7513888° −73.967777°

Like all coordinate systems it can be very accurate. For example at the New York Yankees Baseball Stadium the lat/long coordinates for these two positions are:

- Pitcher   40°49'**36.96**" N   73°55'**40.56**" W
- Batter   40°49'**37.00**" N   73°55'**41.47**" W

Map datum again!
There is no one agreed latitude and longitude map datum (mathematical model of the earth) and consequently different systems of latitude and longitude in common use today can vary by up to 200 m for the same location. This map shows three points which all have the same latitude and longitude using three different map datum (WGS84, OSGB36 and ED50). So long as everyone knows which map datum you are using it does not matter but it stresses the importance of *sharing this information.*

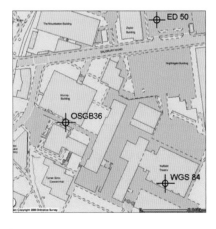

Universal Transverse Mercator (UTM)
After the latitude/longitude coordinate system, the UTM is the most commonly used grid system in the world and many countries, such as Germany and Italy, use it as standard on their national topographic mapping. It differs from lat/long in several ways and is perhaps more straightforward to use.

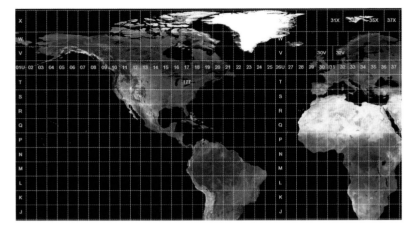

Instead of being a single map projection UTM employs a series of 60 zones, each 6° of longitude wide. The central vertical of each zone (the UTM north/south grid line) is aligned with the central meridian (a line of longitude). Therefore, there is no difference between grid north and true north in the middle of the map. The further you move from this grid line east or west, the greater the difference between grid north and true north becomes – although spatial and directional distortion remains less than one in a thousand inside each zone. The system does not map the polar regions (latitudes greater than 84°N and 80°S), thus avoiding the area where the lines of longitude converge and meet.

- Zones are logically numbered from left to right 1 to 60
- each zone is separated into 20 latitude bands
- they are measured using the metric system (km and m)
- grid values increase from left to right and from the bottom up
- grid references are very logical to follow and read in common with all other grid systems: **Easting** – left to right; **Northing** – bottom to top.

To learn how to read a UTM grid reference see: **Reading a Grid Reference** (pp. 137–8).

As UTM is not a single map projection but a series of independent map projections, one for each of the sixty zones, grids are square to each map but don't match at the edges when joined to a map of an adjacent zone – so for uses such as maritime and air travel, where navigation is over great distances, lat/long is preferred.

Universal Polar Stereographic (UPS)

The UTM map projection is designed to cover the region between latitudes 84°N and 80°S. North and south of this limit (at the poles) another grid system called the UPS is used. At first the coordinates appear be quite cluttered – with different letters and numbers across the grids – yet, like all grid systems, it is actually very straightforward to use once you understand the notation. There is one projection for the Arctic and another for the Antarctic.

- UPS is a square-grid system centred on the poles
- The poles are assigned an imaginary easting and northing of 2,000 km.
- east to west is defined by the 090°W to 090°E lines of Longitude.
- north to south is defined by 000° to 180°

Military Grid Reference System

The Military Grid Reference System covers the whole globe and is derived from the UTM and the UPS grid systems but uses a different labelling system. It is the geocoordinate standard used by NATO and the United States military. Like all grid reference systems it describes an area within a grid, and grid reference sizes vary considerably depending the need for accuracy. So for example, moving a naval fleet may only require 5 letters and numbers, whereas directing smart munitions may require 15.

An example of an MGRS coordinate for Honolulu, Hawaii would be **4QFJ19285736** which consists of four parts:

• **4Q** – is the grid zone designator (GZD)
• **FJ** – is the 100,000m2 identifier

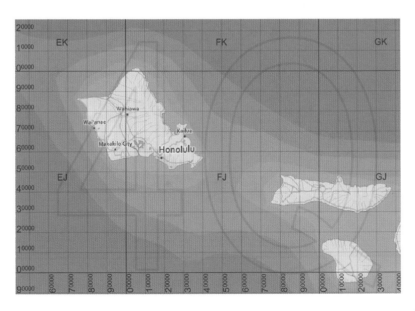

- **1928** – is the easting
- **5736** – is the northing

In this case the MGRS coordinate specifies a location within 10 m x 10 m.

Regional mapping systems

Global mapping systems project large areas of the curved surface of the earth onto flat surfaces – correspondingly, the level of spatial distortion across relatively small areas of the earth can be significant. Furthermore, calculations to relate latitude and longitude to positions on a map can become quite involved. As a result, many countries have developed their own local rectangular grid systems to reduce this degree of distortion – and coordinates can be designated merely by which grid box they are in.

Examples of these are:

- the British National Grid (NG)
- the United States National Grid System (USNG) – derived from the Military Grid Reference System and therefore practically identical in format
- the Swedish SWEREF99 (Swedish Reference System).

British National Grid

The British National Grid reference system is based on the OSGB36 TM map datum (Ordnance Survey Great Britain 1936) and can be used to accurately pinpoint any location in Great Britain and its outlying islands, including the Isle of Man.

It is a system of squares that are subdivided into progressively smaller squares identified first by letters and then numbers – on British OS 1:25 000 and 1:50 000 maps the grid lines are always 1 km apart. ***By specifying the letters and numbers you describe a location.***

- The squares are 500 x 500 km starting with A in the northwest corner to Z in the southeast corner (excluding the letter I). Theoretically, the system extends far over the Atlantic Ocean and well into Western Europe.
- Five of these squares cover Great Britain. The 'O' square contains a tiny area of the North Riding of Yorkshire, almost all of which lies below mean high tide.
- These 500 x 500 km squares were then subdivided into 100 x 100 km squares using the same system of letters. So the N box looks like this:
- Each of these two letters refers to a specific 100 x 100 km of land. Edinburgh is in the NT square while Inverness in the NH square.
- The squares are further divided into smaller squares by grid lines representing 10 km spacing, each numbered 0–9 from the southwest corner in an easterly (left to right) and northerly (upwards) direction.

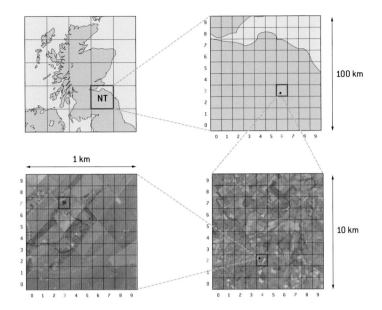

I live inside the grid square NT which covers an area 100 x 100 km. The NT grid further sub-divided – a 6-figure grid reference defines a 100 x 100 m area – my home's location in this case is NT 643 327.

To learn how to read and use grid references see **Using Grid References and Coordinates.**

→ EXPERT TIPS

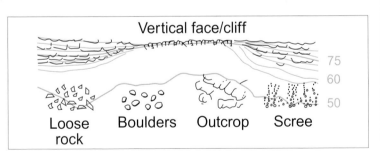

Vertical face/cliff

Loose rock   Boulders   Outcrop   Scree

→ The symbols for cliffs and outcrops of rocks and crag edges can be confusing, so learn them.

→ The Irish Grid currently overlaps the British grid and is used in both Northern Ireland and the Republic of Ireland.

→ A complete printout of OS symbols can be obtained from the Ordnance Survey via **micronavigation.com.**

FEATURES

A feature is anything distinct and fixed that you can see on the land and is invariably marked on the map.

There are four categories of features.

Point features

A clearly identifiable spot on the map which could be large and distant, such as mountain peak used in a **Resection**, or small like a milestone, and used as an **Attack Point**.

Point feature

    Natural point features include: mountain peaks, path intersections, stream junctions and islands.

    Man-made point features – such as communication towers, pylons, lighthouses, trig points and church steeples (see pp. 30–2, **Environmental Navigation**) – have the added advantage that you will often be able to establish the feature's exact height.

Linear features

A continuous edge of any shape which is clearly visible on the map. Most frequently used as **Handrails**, when **Aiming Off** and setting the map. Examples: ridges, walls, streams, tracks, electricity transmission lines, forest edges.

Linear feature

Shape-features

Learning to interpret contour lines on the map and how the actual land looks comes with practice, and the more adept you become at this, the more you will use these 'hidden' clues of the landscape. Examples: valleys, knolls, small depressions in the land, changes of slope angle.

Shape feature

A tussock of grass can be used as a microfeature to take a bearing on.

Micro-features

These are so small they do not appear on your map and are used when no other obvious feature is within view or where these are in line with a larger feature that is likely to disappear as you move, possibly because of contour change.

✖ **QUIRKY FACT:** In 1935 Ordnance Survey started the re-triangulation of Great Britain, an immense task which involved the building of thousands triangulation pillars (trig points) on elevated land, from small hills to mountains. The programme was completed in 1962 and the results were then used to create the British National Grid reference system which would be the basis of the OS's new maps. It also left 6,578 of these iconic landmarks. Many other countries, from Spain to New Zealand, adopted this system and as a consequence have similar pillars which aid your navigation. From any trig point, in clear weather, you should be able to see at least two other trig points.

CONTOURS

A contour is an imaginary line drawn on a map along which all points are at the same height.

The difference between a topographic map and all other maps, including satellite and aerial photographs, is that you can accurately determine the shape of the land using contour lines.

There are many things that can change in a landscape:

- streams and rivers dry up
- paths and tracks are created and can disappear with lack of use
- roads are made and re-routed
- forests are felled and new ones planted
- walls and fences are moved, removed and put up
- electricity transmission lines and communication towers are erected/dismantled
- streams become rivers after heavy rains
- buildings are built and demolished
- entire towns are constructed.

'As part of the advanced navigation course I instruct, I use maps that have been stripped of all information other than contours. The amazing thing is that when the teams are then given a normal map, they complain that it is too cluttered!'

The topography around you is rich with navigational clues: each slope has both direction and shape.

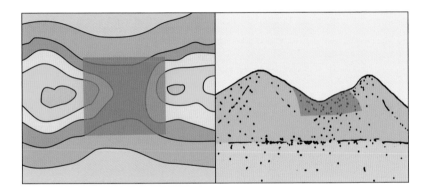

1 **Hill** – A hill is shown on a map by contour lines forming concentric circles. The inside of the smallest closed circle is the hilltop.

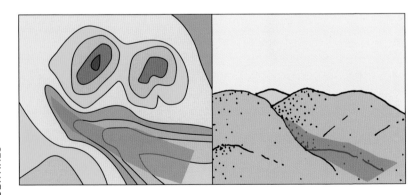

2 **Saddle** – A saddle is normally represented as an hourglass.

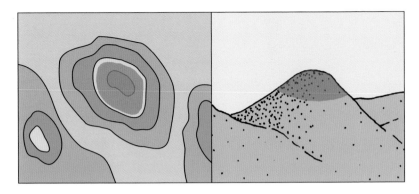

3 **Valley** – The contour lines forming a valley are either U-shaped or V-shaped.

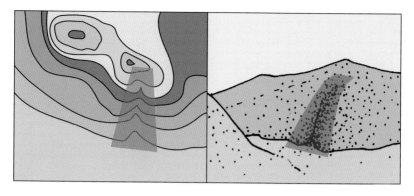

5 **Re-entrant (called a draw in USA)** – The contour lines depicting a re-entrant are U-shaped or V-shaped, pointing ***toward*** high ground.

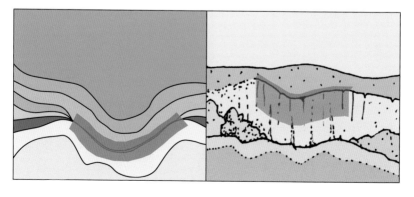

6 **Spur** – Contour lines on a map depict a spur, pointing ***away*** from high ground.

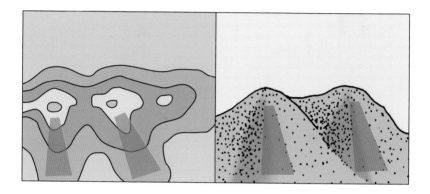

7 **Cliff** – Vertical, or near vertical, slopes are shown by contour lines very close together and, in some instances, touching each other.

Yet the topography of the land – *its shape* – from the hills and mountains to valleys and canyons rarely, if ever, changes. Even after dramatic natural events like a tsunami or devastating man-made events, such as acts of war, the underlying topography remains unchanged. Moreover, two mountains are never identical and even small terrain features (such as dips or depressions) are seldom the same. ***Contours are the single most important detail on your map.***

Each contour is drawn as a continuous, irregularly shaped closed loop and joins land at that specific height above sea level on the map. If you continued to walk the same contour line you would eventually end up where you started – even if that involved a 900-mile walk! They show us not only the shape of the land but also how steep it is.

The ability to look at the shape of the ground (terrain) around you and then identify it on the map and *vice versa* is called ***terrain association***. Learning to recognise the forms illustrated by different contour shapes is a key skill and allows you to identify whether a terrain feature on the map is a cliff or a gentle slope, a hill or a valley.

Practise this continually whenever you are out with the map. Study the shape of the terrain around you and try to imagine what it looks like on the map, then confirm by looking at the map; do this in reverse by looking at the map first. This is such a powerful skill to acquire and the best navigators I know are always excellent at this. The ability to glance at the map and quickly relate it to the surrounding terrain helps you move freely over the land. A major advantage of this technique is the ability to predict in advance what the land that you will be travelling through is going to look like.

At home, if you own **Digital Mapping** that has a 3D function, you can practise this technique at your desk!

→ EXPERT TIP

→ Map contour intervals can change: on OS Landranger 1:25 000 series where the change in height is minimal, often in lowland and coastal regions, the interval becomes 5 m instead of the normal 10 m, so always read the numbers on the contour lines. The same applies to USGS maps.

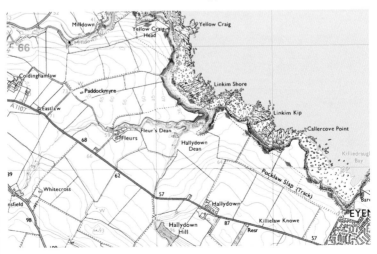

Ascertain your exact height from the contours.

Navigate following terrain features.

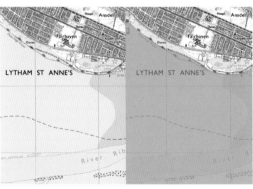

Decide where and when it is safe to walk near tidal water.

Learn your contours – they will allow you to:

- determine your position on a slope
- navigate following terrain features
- travel at the same height rather than over a hill to save energy
- orient the map to match your surroundings
- find your location when you are lost either in the hills and mountains or at the shoreline
- decide where and when it is safe to walk near tidal water
- learn to ascertain your exact height from the contours.

The foreshortening effect

One of the most important concepts to grasp is that because contours represent 3D slopes on your 2D map they betray distance. The distance you would appear to cover on the map is in reality shorter that the actual distance you travel.

Moving up or down any contour on the map will always be a greater distance than you read on your map.

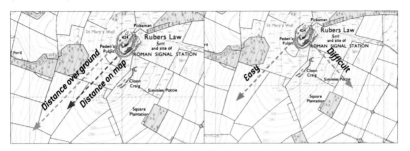

See pages 150 and 155 for tables and calculations relating to this.

There are three terms you will need to learn to be able to successfully use contours in important navigational techniques.

- Contours show the gradient, both up and down – this is called the **Slope**.
- The spacing between the contours represents the **Slope Angle**. The further apart they are, the gentler the slope and lower the slope angle. The closer together they are, the steeper the slope and higher the slope angle.
- The direction a slope faces is called the **Aspect**. So, for example, a slope on the Scottish Highlands facing towards the Atlantic is described as having a westerly slope.

The following photograph is of a knoll (hillock) you encounter as you come off Skelfhill Pen (532 m) east towards Grey Pen (489 m). Up until recently OS did not

display this 9 m-high feature and during the bad Scottish winter of 2009 I was descending from the peak of Skelfhill with Dave 'Heavy' Whalley in a snow white-out. It was tough going and we thought we had started to ascend to Grey Pen – by terrain association ... the land was rising. Since the weather was so terrible we had been using **Pacing** and **Timing** and this conflicted with the location of Grey Pen on the map by 160 m. We stopped, backtracked to our last known fix which was only 20 m away and confident of our position, we retraced the route and walked over the knoll. Even for two experienced navigators (Heavy was Team Leader of the RAF MRT at Kinloss) it was a disconcerting moment.

If the contour spacing is 10 m then features which are less than this will probably not be shown on your map. During daylight, such anomalies are easily recognised – in poor visibility, either at night or in bad weather, they can easily deceive you. If the terrain you are following unexpectedly changes, stop and reconfirm your current location, or backtrack to where you can confidently do so.

Remember features by their shape and not their colour – the latter characteristic can be seasonal!

✕ **EXPERT FACTS:** On OS maps, 1:25 000 scale, contours are spaced every 10 m change in elevation.

➜ On a USGS 7.5 minute map, the interval is 40 ft – unless otherwise stated.

➜ The height above sea level is marked on each contour line.

➜ Every fifth contour line is thicker and called the *Index Contour*; therefore on the OS map they represent a 50 m change in height and on the USGS map 200 ft.

➜ Where the slope is greater than 30° only the index contours are shown.

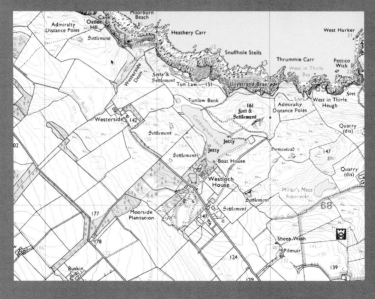

THE COMPASS

The most popular tool to determine north and the degrees of the azimuth is the magnetic compass. They are cheap, durable, dependable, require no power source and are very easy to use.

Introduction

The main uses of a magnetic compass:

- identifying features you can see on the land, by taking their bearing, and finding on your map
- identifying features you can see on your map, using their bearing, and finding them on the land
- taking a bearing from your map and then walking on this bearing
- taking a bearing on a feature you can see and then walking on this bearing
- orientating the map so it matches the features on the land allowing you to identify features around you and check the direction of linear features
- determining your position using bearings from two or more distant features.

When greater accuracy is required get other people in the party to take the bearing and compare results, or use a sighting compass (mirror or prism). The compass binoculars detailed in this section are the most accurate of all devices.

→ BEARINGS: AN INTRODUCTION

A **Bearing** is simply a means of describing the direction between one object and another. Describing my house as being behind me from where I am standing, the car in front or my friend to my right are all bearings. But how do you know where my left or right is if you can't see me and as I move it changes.

Fortunately the earth has a natural magnetic field and a compass detects this and points to magnetic North. From this we can reference the other cardinals.

Accordingly in navigation we use a system which we can all relate to: the cardinals of the compass: N, E, S and W. These cardinals of the compass can be further subdivided (see below right).

Consequently, if I now describe my house as being northeast of me, no matter which way I am facing, it will always be in the same direction from where I am.

This is further refined using an **Azimuth**, the graduated degrees which run clockwise from 0–360° on a compass, where:

North	=	000°
East	=	090°
South	=	180°
West	=	270°

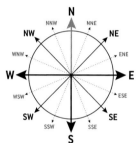

Using this system we can describe where objects are in relation to where we are very accurately.

Bearings, Heading and Course

- **Bearing** – the direction from your current location to a destination
- **Heading** – the direction in which you are actually going
- **Course** – the direction from your starting location to a destination.

✗ In this diagram the initial bearing to the hut is 30° and this is the direct course to it. If you head in a straight line to the hut your course, bearing and heading are all the same. If you stray off this course you will be heading in a different direction and your bearing to the hut will change.

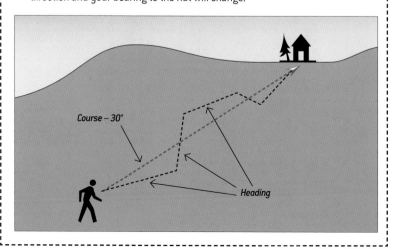

Course – 30°

Heading

→ An easy mnemonic to help remember the main points of the compass is **Never Eat Shredded Wheat**.

→ The terms ahead, behind, left and right are relative to the way you are facing and therefore can be misinterpreted. It is much better if for example you want someone to look in a particular direction to use the cardinals and say 'looking south' or it is 'west of us'.

Very accurate compasses have been invented for determining north which do not depend on the earth's magnetic field and are therefore unaffected by deviation. In vehicles, these are often gyrocompasses or astrocompasses. One which the land navigator will encounter is the fibre optic gyrocompass, found in satnavs and more commonly called an electronic compass. How to use these is dealt with in the **GNSS** section.

How magnetic compasses work

A magnetic compass is a navigational tool for determining direction relative to the earth's magnetic poles. It contains a magnetised needle, which rotates freely on a pivot inside a sealed capsule of liquid (this stops it from swinging violently about), the north-seeking end of this needle is red and always points to magnetic north.

Around this capsule is a moveable bezel which is marked with the main cardinals – N, E, S, W – and divided into degrees. When this bezel is rotated to line up the north marked on it with the red tip of the needle then all of the other cardinals are accurate positioned too. So south is south (180°), west is west (270°).

There are numerous types of compass, from simple fixed dials to military lensatic compasses. This manual uses only one type in land navigation: the baseplate compass and for some specialist terrain, a variation upon this, the Mirror baseplate compass.

→ COMPASS UNITS OF MEASURE

Some countries and organisations developed different units to measure the circle of a compass – most now use degrees. But not all!

Country/Organisation	Units	Designation	No. in a full circle
Global Universal System	Degrees	°	360
NATO	Mils	₥	6,400
Swedish Armed Forces	Mils	₥	6,300
Former Warsaw Pact	Mils	₥	6,000 (anti-clockwise)
Germany	Neugrad	g	400
French Artillery	Grad	G	400
Surveying	Gon	G	400

The baseplate compass

These are the modern mainstay for land navigation and the Suunto M3 Global is used throughout this book. They are essentially a combination of a compass and a protractor.

The type of compass used in this manual has a transparent protractor baseplate for measuring the bearing. It has a direction of travel arrow which shows the direction to follow when walking on a bearing and is used to point in the desired direction when setting a bearing from the map. The protractor baseplate may be rotated with respect to the compass housing for setting bearings.

① The baseplate
The mounting of the compass with anti-slip rubber pads, a ruler in centimetres plus scale rulers for measuring distance on map sizes 1:50 000, 1:25 000, 1:20 000 and 1:15 000.

② Luminous compass needle
A one-zone magnetic needle that pivots on a jewel bearing and is surrounded by liquid so it can rotate freely and smoothly: the **red** end always point to magnetic north.

③ Luminous serrated bezel ring
A rotating bezel which has the cardinal points and the azimuth of 360° marked on it in 2° increments.

④ Orienting lines
Rotate with the bezel and designed to be aligned with the map grid (the blue lines that run top to bottom of your map).

⑤ Luminous base bars
Need to be aligned with the compass needle (red north) to take a bearing (which can then be read on the index triangle).

⑥ Luminous index triangle
This is where you read the bearing.

⑦ Direction of travel arrow
Shows the direction that you want to travel along or the bearing you are taking.

⑧ Adjustable correction scale
Can be set to automatically correct for magnetic declination.

⑨ Stencil holes
Allows you to accurately make marks on your map with a grease pencil.

⑩ Magnifying lens

With practice it is possible to take bearings of features using a baseplate compass to an accuracy of ±4°.

M3 mirror baseplate compass

Designed for use by professionals such as surveyors and architects. They are referred to in this manual in the context of specialist environments, for instance in desert terrain. The reason that they are not my first choice is that they mist up easily, are more fragile than the baseplate and more expensive. However, due to their accuracy, ±1° to 2°, I use them in areas where great precision is required.

A Suunto mirror baseplate compass

Sighting baseplate compass

These compasses are excellent to use in arid climates – rain impairs their performance. They are similar in accuracy to mirror baseplate compasses.

A Brunton sighting baseplate compass

→ EXPERT TIP

→ Small bubbles in the liquid are of no importance. They may appear and disappear with changes of temperature and air pressure.

Compass binoculars

For general navigation these would be a luxury; having said this, keen bird watchers and lovers of nature can justify them. For navigating in some environments, deserts and arctic regions in particular, they are indispensable. Additionally in SAR they have a combined use both as a navigational tool and in visually searching large areas of land.

The main expense in binoculars are the optics, and the cheaper they are the poorer your view will be, which does seem to defeat the purpose of owning them! So if you are going to buy a pair, get the best you can afford.

I use compass Steiner Commander 7x50XP binoculars: spec includes:

- a compass with ten times the magnetic force of conventional compasses, making it accurate to accuracy of ±0.5° to 1°, it is stabilised for quick movement with a built-in illuminator for taking bearings at night.
- Inbuilt reticule for calculating size and distance of objects in the field.
- Lightweight, tough and waterproof so can be used in any environment.
- Nitrogen filled so are condensation-free during changing temperatures and in altitude differences.
- Night Hunter optics for low-light vision, ideal for dusk and dawn navigation as well as poor weather.
- Operating temperatures –40° C to +80° C (I have successfully used them at -55° C).

Thumbs held to the side of the eyepieces helps to eliminate unwanted sidelight.

Setting up your binoculars

You need to set the binoculars for your own inter-pupil distance (PD) – most people's pupils are between 60 and 70 mm apart.

1 Close the binoculars to their lowest eye-distance.

2 Choose an image, sign or a tree at least 100 m away to view. Slowly start to open the binoculars out keeping the vertical reticule upright.

3 When you just pass the maximum view of field slowly push the binoculars together again until your view is unobstructed.

Auto-Focus: the big advantage to this feature is that, once set to your eyes, it allows you to view images from 20 m to infinity without having to refocus – it also works under low-level light conditions, offering the greatest depth of field (difficult to achieve manually).

To set up the auto-focus:

1 Cover one of the objective lenses with your hand or the objective cover. Look through the binocular with both eyes open.

2 On the side not covered, turn the ocular's dioptre setting ring until the image appears clear and sharp. Uncover the lens and repeat the procedure with the opposite lens.

3 When viewing through both lenses, all images from 20 yds to infinity will be bright, clear and in focus.

Make a note of this and your PD settings as they are unique to you, so you can reset the binoculars quickly if required.

If you wear glasses (spectacles).

1 You can choose not to wear glasses when using the binoculars to get a better field of view (as your eye can get closer to the binocular lens), adjust the eyepieces as above irrespective whether you are long or short sighted (possible adjustments of +/– 5 diopters).

② If you have to wear your glasses, there is no need to adjust the diopter settings on the eyepieces, instead fold down the rubber eyecups to reduce the limitation of the field of view.

The Correct way to hold and use binoculars

I first saw this method of holding and sighting with binoculars while working with a NATO tank division in western Germany before the collapse of the Berlin wall – the results of this correct use are dramatic.

To view an object:

① Make sure you have a sure footing (it is easy to lose balance when viewing through binoculars). Remove the lens caps to hang down – not up – from the lens.

② Firmly grasp the binoculars with both hands. Look directly at the object and lift the binoculars to your eyes, without taking your eyes off it. (This stops you unnecessarily searching for the object through the lens).

③ Move your thumbs to the side of the eye-pieces to eliminate any unwanted sidelight – your view will be brighter and clearer.

→ EXPERT TIPS

→ Always replace the lens covers.

→ Steiner compasses are preset for the magnetic field in the northern hemisphere (Zone 2); if you are travelling to another zone you will need to reset them.

4 To take a bearing on the object, place the vertical reticule centre over the object. Holding the binoculars steady, look down to the bottom right and read the object's bearing and glance back at the object to confirm you are locked onto it.

5 The bearing is illuminated by daylight – if the light is low, press the red LED compass light (this will enable you to read the bearing without damaging your night vision).

Determining the distance of objects viewed

If you know the height or length of what you are viewing it is simple to determine the range. Tall structures such as lighthouses are excellent for estimating distance as they are usually highly visible. They can also be used as **Direct Bearings** or for a **Resection** to determine your location. Alternatively, if you are at sea level and can clearly see the top of a hill through the binoculars, possibly a trig point which is clearly marked on your map, you can estimate your distance from it by:

Distance calculated from a known height

Man – height 1.8 m – 10 on the height scale – therefore 1.8/10 x 1,000= 180 m distance. Lighthouse – height 40 m – 20 on the height scale – therefore 40/20 x 1,000 = 2,000 m (2 km).

1 Sight the object with its base on the zero of the height scale on the reticule.

2 Read the scale height at the top of the object.

3 Divide the known height of the object by the scale height and multiply by 1,000.

This is the distance you are from the object.

Distance calculated from a known width

This in effect is exactly the same technique as above only this time you read the horizontal height scale. The use of the vertical scale is preferable (especially on level terrain), since objects are often viewed obliquely along the horizontal axis.

Plateau – width 1500 m – 50 on the width scale.
Therefore 1500/50 x 1,000 = 3,000 m (3 km).

↘ THE THREE NORTHS

You will find references to three types of North in the margin of most maps. At first it appears confusing and many books give lengthy explanations; it is not and they can be explained simply.

True North
- The direction of the North Pole.
- Marked in the skies by the position of Polaris – also called the North Star.
- The imaginary axis upon which the earth spins.
- The point furthest from the equator making it the highest latitude (90°N).
- Where all lines of longitude meet.
- Located in the middle of the Arctic Ocean.
- Sometimes called geographic north.

Magnetic North
- The direction in which a compass needle points.
- Located west of Greenland.
- Slowly moving across the Arctic Ocean – over the last century magnetic north has moved 1,100 kilometres.
- The difference between true north and magnetic north is slowly changing.

Grid North
You can see that the lines of longitude do not run parallel, instead they converge the further north or south from the equator you go. To produce rectangular or square maps, grids have been created and the vertical grid lines on these maps point to what is called grid north. Specifically this is the direction of a grid line which is parallel to the central meridian on the grid.

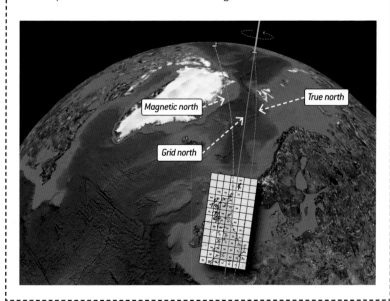

Terms used with compasses

Like all disciplines, navigation has its own language which at first can be daunting – nowhere more so than with the compass. However, there are only three essential, technical compass terms that you need to be aquainted with.

DEVIATION – *false bearing due to local magnetism, natural or man-made (a belt buckle can cause deviation).*

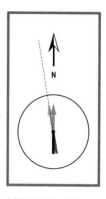

INCLINATION – the needle dipping up and down as magnetic field varies in intensity.

DECLINATION – *difference between true north and magnetic north due to geographical distance between magnetic pole and true pole.*

Declination
Magnetic declination is the angle between magnetic north (the direction the north end of a compass needle points) and true north. When using a map and compass you must adjust for this difference. This is explained in greater detail in the **Techniques** section (see pp. 128–31).

Inclination
The earth's magnetic field changes vertically in intensity and direction and this is called the inclination. These horizontal and vertical components of the earth's magnetic field vary considerably, from vertical at the magnetic poles to horizontal near the

> ✖ **EXPERT FACTS:** The angular differences between the three norths are detailed on most maps. For example, these differences are detailed on UK OS maps in the legend under the heading 'North Points'.
>
> → True north is marked with a line terminating in a five-pointed star on many maps, such as American USGS maps.

> ✖ **QUIRKY FACT:** There is a fourth north called geomagnetic north, which is a theoretical approximation of antipodal points, where the axis of a dipole intersects the earth. It is not used in this manual.

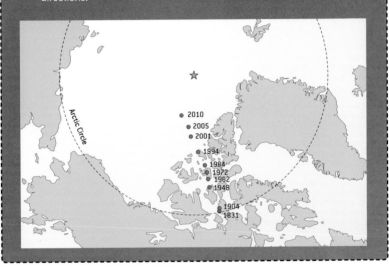

equator. It makes one end of the tip of the compass needle dip down according to the latitude where it is used and therefore drag or stick inside its casing. Due to inclination, compasses must be balanced for different geographical zones in order to keep the needle in a horizontal position.

Suunto have developed a global compass, the M3 Global, which effectively negates this problem so you can use the same compass all over the world. Some manufacturers still divide the globe into separate geographical balancing zones and for each of these regions you need a different compass: check with your manufacturer first.

↘ DECLINATION

When using a map and compass together it is essential to adjust for something called magnetic declination. At first this seems complicated; it isn't and with a little practice you will soon master this technique.

The direction in which a compass needle points is known as magnetic north yet this is not exactly the direction of the North Magnetic Pole. Instead, the compass aligns itself to the local geomagnetic field, which varies over the earth's surface, as well as over time.

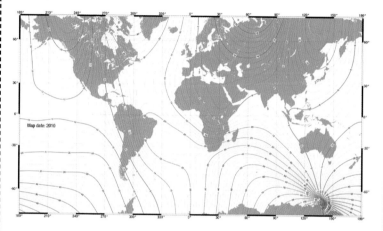

Map date: 2010

The difference between magnetic north and true north at any particular location on the earth's surface is called the magnetic declination.

When we take a bearing with our compass (which aligns itself with the local geomagnetic field) and we want to transfer this to our map (which uses true north) we need to allow for this difference.

Likewise, if we take a bearing from our map and want to transfer it to our compass, again we need to adjust for the magnetic declination of the area.

For example, in the USA the magnetic declination on Mount Rainier, Washington State, in the far west of the country is 16° E whereas in Portland, Maine, on the east coast, it is 16° W. If a compass were adjusted on Mount Rainier and then used in Portland without being adjusted, the error would be 32°! *We must always be aware of magnetic declination.*

At the time of writing, the change as you move across the breadth of mainland Britain was 4°. To see just how easy it is to adjust for declination, go to **Adjusting for Magnetic Declination** section, pp. 128–31.

Deviation

In addition to declination and inclination (which result from the global position and shape of earth's magnetic field) there are other external influences which can cause the compass needle to move away from magnetic north. There are natural and man-made sources for this magnetic deviation.

Natural environmental

Some areas, such as the Cullin Ridge on the Isle of Skye in Scotland, or Ross Island in McMurdo Sound, off the coast of Antarctica, contain large, local iron-ore deposits that make the compass readings inaccurate.

Man-made environmental

Local features that are visible, such as wire fences, railway lines, overhead high-voltage cables, and those which are not visible, including underground pipelines, bunkers, silos and spoil heaps, can all cause magnetic deviation.

Man-made personal

Mobile phones can potentially deviate your compass's needle and, at worst, permanently reverse its polarity. Keep phones and communications radio handsets (SAR team members), steel wristwatches, metal-rimmed glasses, pocket knives, belt buckles or bras (if it is underpinned with wire) away from your compass when in use.

When travelling in a vehicle make sure your compass is not in your pocket next to a door speaker or windscreen wiper motor – the induction coils on these can damage the compass needle.

If you suspect compass deviation:

1. Take a bearing on your map from your known location to a feature you can also see on the land. Correct the bearing for magnetic declination.

2. Take a compass bearing from where you are to the actual feature.

3. Your reading should differ by no more than ±4° – if it does, repeat the technique using another feature to eliminate human error.

If there is deviation use the compass in your satnav or the methods for determining north as described in the **Celestial Navigation** section of this manual.

CELESTIAL NAVIGATION

Celestial navigation was the first comprehensive direction-finding system, used as long ago as 1200 BC by the Phoenicians to travel across the seas.

Introduction

Today celestial navigation is still an essential component of course-plotting used by all mariners ... and should be used by all land navigators. Military Special Forces routinely employ the celestial navigational techniques described here.

This form of navigation is based on the earth's constant and predictable relationships with the sun, moon, stars and planets. There are some very sophisticated techniques requiring the use of specialist equipment such as sextants and astronomical almanacs. **This manual concentrates upon very simple, yet highly effective techniques which require only the equipment you are already carrying – map, compass, watch, pencil and paper.**

An essential component of safe navigation is an awareness of orientation in relation to the cardinals of the compass. The magnetic compass is, of course, the main way to achieve this, and yet there are many occasions where it may be either impossible to use a magnetic compass – for example, as detailed above, the terrain may contain iron ores which interfere with compass accuracy (deviation) – or not convenient, such as for a SAR dog handler who needs to concentrate on directing their dog ... or if you are simply taking a walk for pleasure. Navigation without reference to a compass will improve your experience allowing you to concentrate upon the scenery. Also, crucially, celestial navigation skills could save your life in an emergency situation where you have lost or broken your compass.

Celestial Navigation is divided into two sections: **Daytime** and **Night-time** navigation. There is a small amount of theory included that will give you a greater understanding of why celestial navigation works and help you to logically think through the techniques you are going to learn.

Estimating angles is an important component of celestial navigation and using your watch face is a great way to simplify this – every 5 minutes = 30°.

The science

Understanding the movement of the earth, sun, moon and planets will greatly help you to know what to:

- look for
- expect to see
- do with this information and how to interpret it.

We all know that it takes the earth one year to orbit the sun. Less well known is that the axis between the North and South poles, upon which the earth spins, is tilted at an angle of 23.5° to the sun. For half the year, the northern hemisphere is tilted away from the sun while the southern hemisphere is tilted towards it, which explains why:

- for half of the year the North Pole is in darkness while at the same time the South Pole has constant daylight. The other half of the year the north is in daylight and the South is in darkness
- when it is summer in the northern hemisphere, it is winter in the southern hemisphere and *vice versa*
- the sun only rises exactly in the east and sets exactly in the west twice a year
- we have seasons
- the sun appears at different positions in the sky throughout the year.

The relative movement of this tilt is marked by special days which are called the solstice and the equinox.

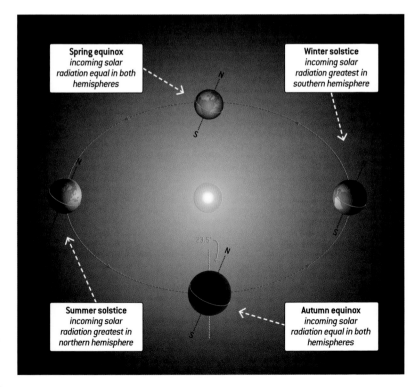

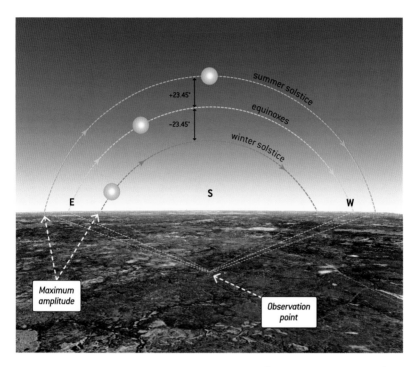

+23.45°

−23.45°

summer solstice

equinoxes

winter solstice

E

S

W

Maximum amplitude

Observation point

The path of the sun from sunrise to sunset varies throughout the year (here in the northern hemisphere). The azimuth where the sun hits the horizon varies as the sun moves from solstice to solstice.

Solstice
Every year there is a Summer Solstice and a Winter Solstice when the Sun's apparent position in the sky reaches its northernmost or southernmost extremes.

Northern Hemisphere Solstices
- Summer Solstice is achieved when the North Pole is tilted closest to the sun, usually 20–21 June, and the period of daylight is at its longest in the entire year.
- Winter Solstice is achieved when the North Pole is tilted furthest from the sun, usually 21–22 December, and the period of daylight is at its shortest in the entire year.

Southern Hemisphere Solstices
- Winter Solstice is achieved when the North Pole is tilted closest to the sun, usually 20–21 June, and the period of daylight is at its shortest in the entire year.
- Summer Solstice is achieved when the North Pole is tilted furthest from the sun, usually 21–22 December, and the period of daylight is at its longest in the entire year.

Equinox
As the earth slowly tilts south, the sun will be directly overhead at the equator for one day only; as the earth slowly tilts back, the sun will again be directly overhead at the equator for one day only: these two equinoxes occur in March and September every year.

Daytime celestial navigation

The sun, which is a star, provides us with a wealth of navigational information from determining the cardinal points to calculating local magnetic declination.

Radial arms

A radial arm is any object which is clearly visible and gives you a reference to the cardinals of the compass. As discussed in the **Environmental Navigation** section communication towers are an excellent example to use, better still are the sun and the moon.

This is the most basic and straightforward of techniques, using the sun as a reference point and one which I constantly employ to maintain a bearing without having to continually refer to my compass.

1 Face the direction (bearing) you are going to travel and reach out with your arm as if you were going to grab hold of the sun and hold this position for a couple of seconds: this helps imprint your orientation to the sun.

2 Lower your arm and move forward, keeping in mind where the sun should be in relation to you.

3 You can travel for 10 mins on a bearing using this technique before you need to repeat it. If the sun is behind you, use your shadow. Reach out and hold your arm parallel with your shadow and hold this position for a couple of seconds.

→ **EXPERT TIP**

→ If you are in a forested area, use your own shadow plus those of vertical tree trunks.

Sun timetable

On any day you can calculate the exact position of the sun in the sky at a specific time and use this to navigate. You can create a sun timetable for anywhere in the world and on any day. This information can be easily downloaded for free from the US Naval Observatory Astronomical Observations Department's website: see **micronavigation.com**. You need to input the date you will be travelling, the time interval you require for the data (hourly is fine) and the location in longitude and latitude.

The following example was created for Los Angeles on the 6 October 2009. Anywhere in the city at 06:00hrs the sun will be almost due east of where you are and at 14:00hrs it will be southwest. By knowing these bearings during the day, you can easily determine the other cardinals/bearings of the compass.

LOCATION: Los Angeles   **LAT/LONG:** W118°22", N34°05"   **DATE:** 6 Oct 2009

TIME	ALTITUDE (°)	BEARING (°)	TIME	ALTITUDE (°)	BEARING (°)
06:00	1	97	12:00	50	187
07:00	13	106	13:00	46	209
08:00	25	116	14:00	39	226
09:00	35	128	15:00	29	240
10:00	44	144	16:00	17	250
11:00	49	164	17:00	6	260

Note: local time used. Altitude refers to the height of the sun above the horizon in degrees. The compass bearing of the sun is referred to as the azimuth.

The table you compute is accurate for a large area. From my home, I can travel 100 km east or west before the values change by 1° and 500 km north or south for the same change. In addition, they remain precise enough for general navigational purposes for 10 days. These bearings refer to azimuth (true bearing) to which magnetic declination needs to be applied to get the equivalent magnetic compass bearing.

So applied to our example: declination at Los Angeles is 13° E. So at 1600 when the azimuth of the sun was 250°, its magnetic bearing was 237° (250° minus 13°).

→ EXPERT TIPS

- → Print your sun timetable out onto waterproof paper and carry it with you in a safe place.

- → An easy to remember mnemonic for declination is: 'Declination east, compass reads least. Declination west, compass reads best'.

- → You can work the timetable in reverse and tell the time by taking a bearing on the sun and checking this against the corresponding time on your table.

THE ESSENTIALS

Finding north/south using a watch

There may be an extreme case when you do not have a compass, spare compass or satnav, or that, for whatever reason, none of them are functioning correctly – you can still find the cardinals of the compass.

Using your analogue watch to find south in the Northern Hemisphere

- Point the hour hand directly at the sun.
- Bisect the angle between it and the 12 o'clock mark.
- This direction is south.

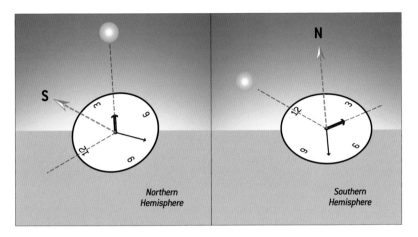

Northern Hemisphere

Southern Hemisphere

Using your analogue watch to find south in the Southern Hemisphere

- Point the 12 o'clock mark directly at the sun.
- Bisect the angle between it and the hour hand.
- This direction is north.

If your analogue watch is set to daylight-saving time, bisect the angle between the hour hand and 1 o'clock on the watch face.

→ EXPERT TIPS

- → If you have a digital watch, draw an analogue watch set at the current time on a piece of paper and use this.

- → This method improves in accuracy the further away from the equator you are.

Shadow-tip method using the sun – Northern and Southern Hemispheres

The sun is always moving east to west, in both hemispheres, so the shadow moves from west to east. The first mark will always be west of any subsequent mark.

- Place a stick, walking pole etc., a minimum of 1 m height into the ground.
- Mark the tip of its shadow. Wait minimum of 20 mins.
- Mark the new tip of its shadow.

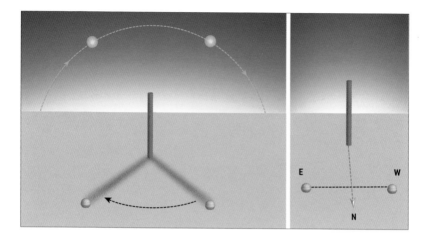

- Draw a line connecting these marks. This line runs west to east from your first mark.

This is a relatively accurate method (±12°) and best performed within 2 hours of the time of day when the sun reaches its maximum altitude – when the shadow is shortest.

A more accurate way (±6°) is to measure the shadow cast sometime before the sun is highest and the shadow shortest (noon). Mark this spot. Wait till the shadow reaches that length again. Mark this second spot and draw a line to the first mark. The line will run approximately west-east.

↘ EXPERT TIPS

→ Close to the equinoxes (21–23 March and 22–23 September), this method is very accurate throughout the whole day.

→ The shadow-tip system cannot be used in polar regions (above 60° latitude in both hemispheres).

→ At midday, the sun is at its highest point – in the Northern Hemisphere it will be to the south at that time; in the Southern Hemisphere to the North. Be aware that midday during Daylight Saving Time will be at 13:00 hrs not 12:00 hrs.

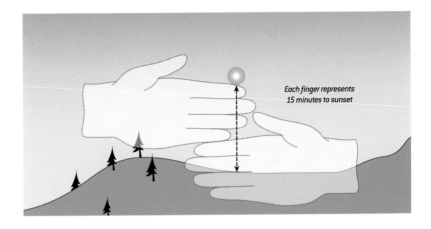

Each finger represents 15 minutes to sunset

Estimating time to sunset

Knowing the times of sunrise and sunset for the area you are navigating allows you estimate how many daylight hours you can travel. However, if you do not know these, you can calculate the hours of daylight left with your hands so you can reach your destination before it gets dark.

To estimate when the sun will disappear behind a hill, ridge, horizon, or canyon floor:

1. Hold both your hands at arm's length, palms facing you and fingers horizontal.

2. Line up the bottom of the sun with the top of the upper finger with one hand, and the other hand lined up to the bottom of the other hand.

3. Count the number of fingers to the horizon. Every finger is about 15 mins of daylight left before sunset. You can estimate up to two hours before sunset with this method using both hands.

Special Forces on deployment.

Night-time celestial navigation

Radial Arms
This is performed in exactly the same way as during the day, only using the moon.

1 Face the direction (bearing) you are going to travel and reach out with your arm as if you were going to grab hold of the moon and hold this position for a couple of seconds: this helps imprint your orientation to the moon.

2 Lower your arm and move forward, keeping in mind where the moon should be.

3 You can travel for 10 mins on a bearing before you would need to repeat this technique, because the moon would have moved sufficiently by then.

Finding direction from any star
If your view of the sky is limited and you are unable to see the circumpolar constellations (see pp. 80–5) you can still find an approximate direction by choosing any bright star mid-sky (avoid stars near the poles) and noting its position relative to a peak, or the horizon (or by sighting along a stick, rock etc.). This technique is not hemisphere dependant – it works exactly the same in both hemispheres. Wait 10–15 mins and then note in which direction the star has moved:

- Up – in the east • Down – in the west • Left – in the north • Right – in the south

Moon timetable

Just as you can calculate the exact position of the sun in the sky at a specific time during the day you can also do the same for the moon. The computations are more complex because the Moon makes a complete orbit around the Earth with respect to the fixed stars about once every 27.3 days (its sidereal period). However, since the Earth is moving in its orbit about the Sun at the same time, it takes slightly longer for the Moon to show the same phase to Earth, which is about every 29.5 days (its synodic period). Visit **micronavigation.com** to find the US Navy charts for your area.

You will need to input the date you will be travelling, the time interval you require (hourly is fine) and the location in longitude and latitude.

The following example was created for Los Angeles on the 6 October 2009. Anywhere in the city at midnight the moon will be due east of where you are and just before 07:00hrs it will be due west. By knowing these bearings during the night, you can easily determine the other cardinals/bearings of the compass.

LOCATION: Los Angeles   **LAT/LONG:** W118°22", N34°05"   **DATE:** 6 Oct 2009

TIME	ALTITUDE (°)	BEARING (°)	TIME	ALTITUDE (°)	BEARING (°)
21:00	4	62	03:00	81	81
22:00	15	69	04:00	75	75
23:00	27	76	05:00	64	64
00:00	50	91	06:00	52	52
01:00	62	101	07:00	40	40
02:00	73	119	08:00	28	28

Note local time used. Altitude refers to the height of the moon above the horizon in degrees. Figures rounded up to nearest whole

The table you compute is accurate to the same degree as the sun tables oulined on page 73. These bearings refer to azimuth (true bearing), to which magnetic declination needs to be applied to get the equivalent magnetic compass bearing.

Shadow-tip method using the moon
The moon is always moving east to west, in both hemispheres, so its shadow moves from west to east. The first mark will be west of any subsequent mark.

1 Insert a stick or walking pole into the ground at a minimum height of 1 m.

2 Mark the tip of its shadow. Wait a minimum of 20 mins.

3 Mark the new tip of its shadow. Draw a line connecting these marks.

4 This line runs west to east from your first mark.

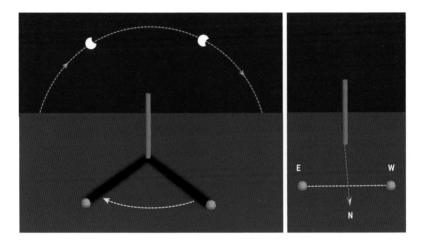

↘ EXPERT TIPS

→ The shadow-tip system is not intended for use in polar regions (above 60° latitude in both hemispheres).

This is a relatively accurate method (±12°) and best performed within 2 hours of the time of day when the moon reaches its maximum altitude – when the shadow is at it's shortest.

A more accurate way (±6°) is to measure the shadow cast sometime before the moon is highest and the shadow shortest (meridian passage). Mark this spot. Wait till the shadow reaches that length again. Mark this second spot and draw a line to the first mark. The line will run approximately west–east.

Constellations in the northern hemisphere

A constellation is a group of stars that form a pattern or shape we can make out in the night sky.

The North Star

This star is unique in the night sky as it sits almost exactly at celestial north pole – it is also called Polaris.

Unlike any other star, the sun or the moon, the North Star is always in the same position in the sky and therefore offers an excellent way of determining the cardinals of the compass at night. One advantage is that the North Star, though only second in magnitude (brightness), is all by itself in a black patch of sky. If you focus on the patch, your eye will automatically fix on the North Star.

As it is actually not a particularly bright star, we also use one or more of the five circumpolar constellations to signpost our way to it. These are:

• The Big Dipper/The Plough (Ursa Major) • The Little Dipper (Ursa Minor) • The King (Cephus) • The Queen (Cassiopeia) • The Dragon (Draco)

These groups of stars always rotate anti-clockwise around the North Star and never set, meaning that they are always visible on a clear night.

Finding the North Star

❶ Find and face The Big Dipper. Draw an imaginary line from the so-called 'pointer stars' that form the leading edge of the dipper.

❷ Continue in a straight line, about five times the distance between these two stars until you get to the next bright star (it will be slightly right of your imaginary line); you have found the North Star.

❸ Confirm this is the North Star by finding The Queen which is opposite to the Big Dipper (it has five stars that form a shape like a 'W' on its side).

❹ Point up to it with both arms outstretched, hands together. Slowly lower your locked arms to the horizon – this is true north.

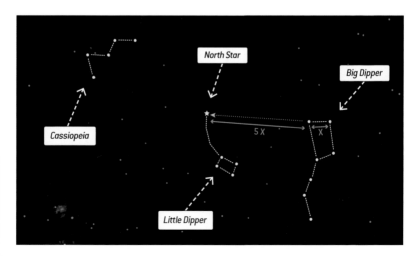

At lower north latitudes the North Star can sometimes be below the horizon. Then Cassiopeia, since it is on the other side of the pole, will be up. An imaginary line that bisects the angle between the more open of the two V's comprising the letter W of Cassiopeia, points to Polaris.

Determining magnetic declination using the North Star

In the northern hemisphere, if you can see the North Star, the declination can be determined as the difference between the magnetic bearing and a visual bearing on the polestar. Polaris currently traces a circle 0.75° in radius around the north celestial pole, so this technique is accurate to within a degree.

At high latitudes a plumb-bob is helpful to sight Polaris against a reference object close to the horizon, from which its bearing can be taken.

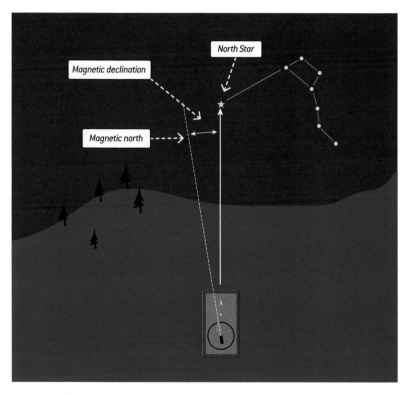

↘ EXPERT TIPS

→ Planets in our solar system are not used for this type of navigation yet reflect the sun's light and can look like stars, the way to tell the difference is that they do not twinkle – really – the twinkle is the result of particles interfering with the light from the star as it travels millions/billions of kilometres.

Latitude from the North Star

The number of degrees that the North Star lies above an unimpeded horizon is approximately the same as the number of degrees of your latitude.

1 Hold your arm out in front of you.

2 The width of your flat hand from the outside of your thumb to your little finger is approximated 10°.

3 Estimate how many 'hands' there are between the horizon and Polaris. This is very approximately your latitude north.

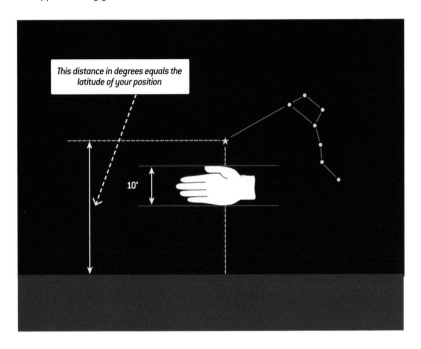

This distance in degrees equals the latitude of your position

10°

↘ EXPERT TIPS

→ If you find it difficult to visualise the degrees, draw a compass bezel and mark the cardinal degrees.

→ There are three simple ways to determine location in latitude and longitude:
  • Google Earth: type in the location, or move the cursor, over the area you will be navigating and read off the location in lat/long.
  • Satnav: set the unit to read lat/long and move your pointer to the location on your screen.
  • Map: in addition to regional mapping systems many maps have lat/long marked on them – see how to read in the **Maps** section.

Constellations in the southern hemisphere

The North Star is not visible anywhere south of the equator. In addition, there is no equivalent of it in the night sky, so determining south is a little more complicated. Fortunately, there is the Southern Cross and there are four first magnitude (brightest) circumpolar stars. From latitude 33° S southwards, the cross is above the horizon at all times.

Finding due south

There are three ways to find due south (also called celestial south) using the Southern Cross constellation.

Finding the Southern Cross for the first time can be tricky as it is relatively small and looks more like a kite than a cross. In addition, there are a few other constellations which look like crosses. Fortunately, Rigel Kentaurus (Alpha Centauri) and Hadar (Beta Centauri), the pointer stars, help to confirm that you have located it. No other cross has a similar signpost.

1 Draw an imaginary line between the bright stars Acrux and Achernar.

- Due south is halfway along this line.
- Now point your hands, one at Achernar and the other at the closest of the pointer stars to the Southern Cross (Beta Centauri).
- Bring your hands steadily together until they meet. They are pointing at the celestial south pole. Drop your hands straight down to the horizon. That marks due south from where you are.

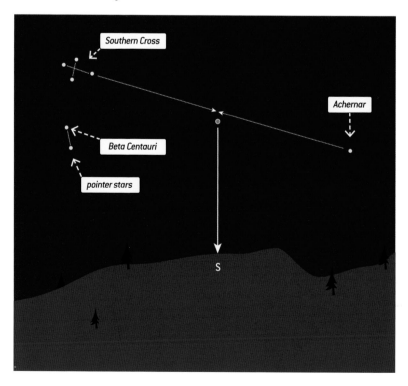

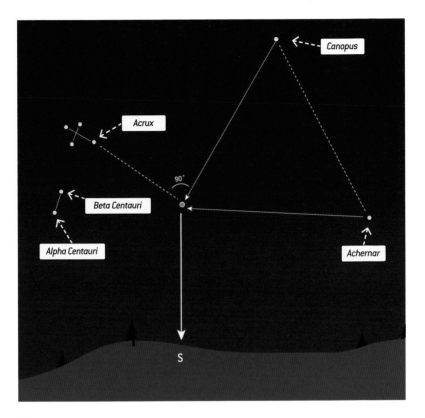

② You can also use the stars Achenar and Canopus.

- Imagine that these two stars are the base of an equilateral triangle.
- The apex is at the celestial celestial south pole.

③ Celestial south is virtually at the point where imaginary lines from Canopus and the upright of the cross meet at a right angle.

- The bottom of the cross is about five lengths of the upright of the Southern Cross from celestial south pole.

↘ EXPERT TIPS

→ The United States Naval Observatory website, see **micronavigation.com**, also gives rise/set and meridian passage (i.e., when they will bear true south for Hadar and Rigel Kentaurus (the Pointers for the Southern Cross) Canopus and Achenar; and Mimosa the brightest star of the Southern Cross, the one on the left of the cross bar when the cross is standing up.

Orion – a constellation for both hemispheres

The one star constellation that can be viewed and used in both hemispheres – Orion! This constellation circles the earth on the celestial equator and consequently always rises in the east and sets in the west, no matter which latitude you see it from in either hemisphere and can be used to very accurately determine east and west.

In northern latitudes in winter and southern in summer Orion's leg is represented by one of the brightest stars in the night sky, Rigel, and from this it is easy to find Orion's Belt, the middle star of which, called Alnilam, is only 1° from the celestial equator.

West

Wait until Orion begins to set and when this middle star touches the horizon this is due west.

East

Watch for the bright reddish star Betelgeuse to rise from below the horizon followed by the Belt, the middle star is due east. Finding east takes practice.

The arc from horizon through Alnilam is called the 'setting angle'. It lets you locate west up to three hours before Alnilam sets, by sweeping an arm down the line it will follow. The angle of this line to the horizontal is 90° minus your latitude. Once you've worked out the angle, you can gauge it by holding up a branch or pole. Even the crudest sort of protractor would be a help here. You can do the same thing to locate east up to three hours after Orion's rising.

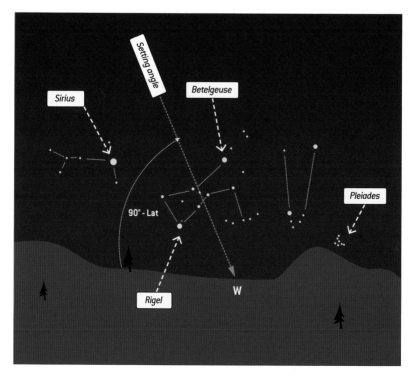

SECTION TWO
TECHNIQUES

LESSON PLANS

I personally recommend that everyone from a novice to an experienced navigator follow these lesson plans.

In just two weekends, the complete beginner can learn the core techniques that underpin all land navigation – giving them the freedom to roam safely. The experienced navigator can consolidate their knowledge, discover best practice and learn newly developed techniques. Over four weekends all the techniques described in this manual for map and compass navigation are covered – this will provide the reader with the skills to become an expert navigator on all terrains and in all conditions.

I have prepared a detailed lesson plan for the first weekend/two-day training. Use this same structure for the subsequent three training phases. Full lesson plans for all four weeks, plus **GNSS and digital mapping** are available as a downloadable file from **micronavigation.com**.

Weekend/Phase ❶ (Beginner)

Saturday (Day 1)

Learning Outcomes
- How to **Orient your Map**
- How to adopt the **Brace Position**
- How to aim for an **Attack Point**
- How to select a **Handrail**
- How to recognise and use **Collecting Features**
- How to **Thumb the Map**

Equipment

- this manual
- a map of the area
- a grease pencil/chinagraph
- some plain paper

Preparation

To prepare for your first stage of learning read these sections first:

- **Introduction** (pp. 12–23)
- **Emergency Calling Procedure** (pp. 351–4)
- **Maps** (pp. 33–45)
- **Contours** (pp. 49–55)

Location and conditions

- An area you know well – municipal parks in towns and cities are ideal, especially for beginners, as the terrain is safe, it is difficult to get lost and detailed maps should be available.
- Good weather conditions – make sure that you are confident you can return to your start at any time.
- Tell somebody where you are going and when you expect to be back and make this routine for every trip you plan and undertake.

Tasks

To learn a new technique simply select the page it is described on and follow the instructions of how to perform it. Each technique described in this manual follows the order in which you should learn them.

1 Orient your map to your surroundings and identify on the map exactly where you are.

2 Get into the **Brace Position**, **Fold your Map** and **Position Mark** your location using your grease pencil.

3 Choose an **Attack Point** on your map – such as an entrance/gate, a building or monument if marked on your map.

4 Select the easiest **Handrail** – such as a path, track, fence or edge of a playing field – to your feature and mark your route to it.

5 Picture in your mind the **Collecting Features** – such as a path intersection, a pond or building you pass en route – you will see as you travel to your attack point and mentally cross them off as you pass them, confirming where you are on the map.

6 Create a **Catching Feature** – such as a stream or a road – beyond your attack point, in case you overshoot your attack point.

7 Lastly, before you set off, mark your thumbnail and, as you hold the map, place this mark next to your current location. As you travel, keep moving your thumb to where you are.

Reinforcement

- Spend time acquiring these new skills
- Choose different Attack Points and use the techniques in isolation and in different combinations

TECHNIQUES

- Refer to the instructions in the manual at first but once you become confident try to do without the manual
- Explain/demonstrate to someone else how to carry out these techniques

Sunday (Day 2)

Learning Outcomes
- How to determine your **Pacing Count**
- How to understand **Timing** over a **Measured Distance**
- How to use **Transit Lines**

Equipment
- The same items as for Day 1
- 50 m paracord (or long tape measure)
- A watch which shows seconds (digital ideal)
- A ruler
- A 1:25 000 map of the area.

Location and conditions
As on Day 1.

Preparation
Go over the six fundamental techniques you learnt on Day 1.

Tasks

1 Take time to measure precisely the marked distance for determining pacing count and ensure it is level ground.

2 Measuring distance on a map requires concentration; always double-check your reading.

3 Initially learn timing on level, easy to cross ground and introduce pacing, in combination, as soon as you can; used together they are remarkably accurate.

4 Transit lines should become routine for you every time you move.

Reinforcement
As day 1 – plus, remember, practice makes perfect! In just two days you will have learned to navigate proficiently and safely using only a map.

Weekend/Phase **2** (Intermediate)

- **Taking a Bearing**
- **Working with Bearings using a Map & Compass**
- **Back Bearings**
- **Triangulation**
- **Drift**
- **Taking a bearings on map and transferring to a compass**
- **Adjusting for Magnetic declination**
- **Baselines**

→ First time navigation – write down the **Collecting Features** you expect to see and physically tick them as you reach/pass them.

→ With many of the experienced SAR navigators that I instruct, we start working in municipal parks, where it is easy and safe to get to grips with even the more difficult and complex techniques.

→ Run through these techniques whenever you can – practice makes perfect!

- **Reading a Grid Reference**
- **Back Snaps**
- **Aiming off**
- **Boxing**
- **Radial arms**

Weekend/Phase ❸ (Advanced)

- **Route Planning**
- **Routes**
- **Slope Aspect**
- **Dead Reckoning**
- **Relocation Procedure**
- **Leapfrogging**
- **Outriggers**
- **Distance – Visually Estimating**
- **Contouring**

Weekend/Phase ❹ (Expert)

- **Cliff Aspect**
- **Bearings on the Move**
- **Warning Bearings**
- **Distance to horizon**
- **Distance off**
- **Searching**
- **Parallel Errors**
- **Working with Grid References**
- **Converting Grid References**

1

ORIENTING THE MAP

All maps are drafted with north at the top – by rotating the map to align with north you will match the position of the features you can see around you to those on the map.

This is a great navigational technique to keep you on track. Once mastered, it can be the primary navigational technique you use on the move – you don't need to stop to do it. The two easiest ways to perform this technique are:

Using your compass

Just put your compass on the map, look at where the compass north (red) is pointing and rotate the map until the needle runs parallel to the vertical grid lines on your map: it does not matter which way the compass is pointing, only the needle!

Orienting the map using the environment

Using your environment

If you know where the sun is (mornings easterly/afternoons westerly), roughly orient your map in relation to the position of the sun and look for features in your surrounding landscape that are marked on your map. Cliffs, paths, walls, streams and coastlines, are excellent linear features to align with.

Terrain Association

In addition to these two techniques, to become a really skilled navigator you need to learn how to orient the map using just the shape of the land alone. This is especially important when navigating across seemingly large, featureless areas, such as open moorland or heath and when navigating where visibility is restricted, either due to the environment, for example in jungles, or due to bad weather or at night. *The shape of the land becomes the feature.* This is a difficult concept to grasp initially as we are used to looking for single objects, such as buildings, or linear features, such as tracks. If they are not present or have changed course, the shape of the land is key to continually reaffirming location when moving across it.

→ If the **Magnetic Declination** is more than 10° you will need to make an adjustment. To do this, set the compass dial to the degrees of magnetic declination east or west, align the edge of the compass with the vertical grid lines, and then rotate the map until the needle runs parallel with the orienting lines in the compass.

→ The edges of the map are oriented true north/south, but the grid lines are tilted by the amount stated on the map. In some areas this could be important.

IMPORTANT
Move around the map to keep it pointing north, as if it was laid on the ground, so no matter which direction you are facing, the land in front of you will also be the land in front of you on the map.

TECHNIQUES

BRACE POSITION

The brace position creates a stable platform to both take precise bearings and when working with the map and compass together.

You should adopt this position as a matter of standard practice.

Taking a bearing from a feature on the land

❶ Turn and face square on the object on which you wish to take a bearing.

❷ Assume the brace position with your working knee facing the feature. Place your compass on your knee pointing to the feature.

→ Adopting the brace position while out with a group signals to the others that you are working and should be left to get on with it.

→ Work with the map facing away from the prevailing wind and weather. If on a steep slope make sure you are parallel with the contour lines for stability.

Occasionally it will not be practical or necessary to assume the brace position, such as when quickly checking the bearing you are following.

1 Standing, turn and face square on to the feature.

2 Lift your compass to just above waist height and away from your body to take the bearing.

FOLDING THE MAP

Most printed maps are large, yet as you will be navigating in just a small part, you only need to display this area.

People are often amused that I spend time teaching them how to fold a map but having seen people with what I call towels blowing in the wind, and given the cost of these items, it is worth taking time to care for your maps.

In fact I recommend displaying only the section you will be navigating for the next half hour, then stopping, reviewing the map and selecting the next area. Too much detail can be confusing and forever having to search a large area of map to relocate is pointless.

In common with most national mapping agencies all OS maps fold the same way with pre-pressed creases. To fold a completely unfolded map:

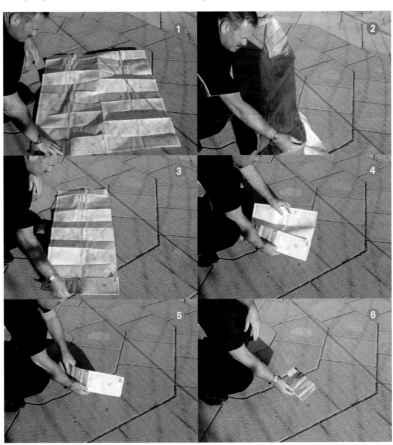

→ If weather conditions are poor, particularly in a strong wind, I use my grease pencil to mark the grid numbers on one easting and northing so that I do not have to completely unfold the map to check the grid reference.

→ Use an elastic band to keep your map folded to your desired area to display.

- Lay flat with the title cover, face up, in the bottom left-hand corner.
- Fold the map in half bringing the top edge to meet the bottom edge.
- Fold from the cover along the creases keeping the title cover face up.
- Fold in half so that the title cover is on the outside.

When you wish to select an area of the map on a 1:25 000 I would work with 12 x 12 cm which is an area 3 x 3 km. The width of the standard creases on OS Landranger and Explorer maps is 12 cm.

POSITION MARKING

Ideally you should use laminated maps. Not only are they much more durable than paper maps, they are waterproof and can be marked with a grease pencil.

Grease pencils/chinagraphs are cheap, create a waterproof mark and when you no longer need the information you have recorded on your map, you can remove the marks without causing any damaging to the map surface.

Always mark your start point and thereafter mark points where you have changed bearing or obtained a fix.

A good idea is to join these marks up as you move so you create a visual track record of where you have been.

ATTACK POINT

To follow your route to your objective you navigate a series of short legs. Each leg starts from a known point and leads to an identifiable point on the map known as an attack point.

In this example, the objective is the cairn (above left). The noticeable tussock of grass is used as a closer attack point on the correct bearing – as we advance the rolling terrain obscures our objective, but we can be confident that we are still on our bearing. This is confirmed when the land drops away and we find our objective where we expected it.

With good visibility your attack point could be in theory as far as the 'eye can see'. However, as you travel, the contours of the land can alter causing you to lose sight of it; visibility can change with anything from a sandstorm to fog.

Generally the nearer an attack point, the safer your navigational leg, so I use a maximum of 500 m. Even if I can clearly see a point kilometres in the distance, I will select another feature closer to me, on this same bearing, and aim for this.

If the terrain or weather/visibility is poor or if you are near areas of danger, such as cliffs, choose much closer attack points, preferably between 10 m and 30 m. In extreme conditions such as a snow blizzard, this can be as little as 5 m.

TECHNIQUES

HANDRAILS

A handrail is an an easily identifiable linear feature, marked on your map that you can follow towards your next attack point.

Typical handrails might be:

- walls
- fences
- stream-beds
- riverbanks
- ridges
- valleys
- paths, roads or tracks
- forest edges
- overhead power lines.

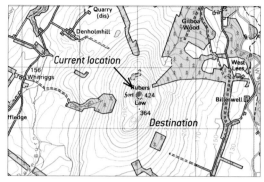

If visibility is reduced, either in poor weather or low light levels, then following a handrail is the safest form of travel – and in severe conditions becomes essential.

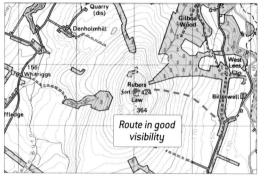

Route in good visibility

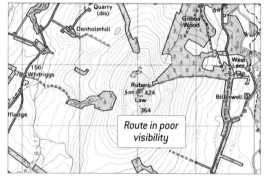

Route in poor visibility

↘ EXPERT TIPS

→ Motorway Exit Syndrome: When following a track edge in poor visibility you can by mistake follow another track that leads off it, so stick to the centre of the track.

→ Errors can be made where there are parallel potential handrails in a landscape with many similar features, for example walls – before you start to follow a handrail check your location first.

→ If following a fence note the condition and age of it; if these change en route stop and re-confirm your bearing as it may be a new section of fencing.

TECHNIQUES

COLLECTING FEATURES

These are features you predict or know will be on your path and mentally collect along the way.

Creating a mental image of what is coming next on your journey is first-class navigation.

As you navigate to your next **Attack Point** you can confirm that you are following the correct track using features which you have identified will form part of your track.

These can be almost any feature marked on a map and are easily categorised into three types, from the most precise (❶ placing you exactly – a fix), to those putting you in the general area (❸ an estimated position – EP).

❶ **Spot features:** such as bridges, intersections of paths, junctions in rivers/streams, cairns, summits.

❷ **Linear features:** such as walls, streams and ridges with no junctions.

❸ **Area features:** the terrain may change from rocky to marshy to rolling; should you be going uphill or downhill, or reaching a level ground or a particular land feature.

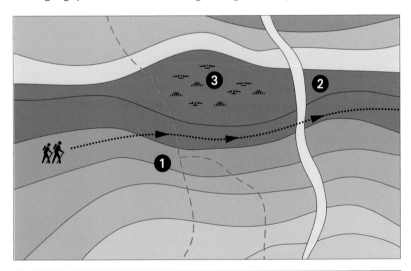

↘ EXPERT TIPS

→ Collecting features and visual **Back Snaps** used together are an excellent way of keeping track of where you are at all times.

→ Have a minimum of five features with which you can confirm your position when in doubt.

CATCHING FEATURES

These are features you predict will be on your path, which if you reach, signal you have overshot your destination.

Ideally this is a linear feature which runs perpendicular beyond your **Attack Point** so if you reach it you know that you have overshot your attack point.

Rivers, streams, walls and roads are ideal. Conversely, it could be the edge of a forest, a lake or a definite change in the **Contours**, such as a depression in the land.

1 Prior to undertaking a critical leg of navigation, study your map and look for a catching feature beyond your intended direction of travel.

2 It should be wide enough to allow you a good margin of error so if you miss your attack point substantially you will still be 'caught'.

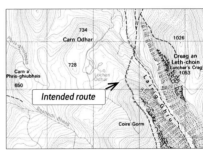

Intended route

3 Stop and take a bearing from a feature which you can see and which is on the map to get a fix of where exactly you are along your catching feature.

4 If there is no obvious feature immediately next to this, pace out in one direction along this linear feature and search for a point where you can take a second bearing to get a fix.

5 Do not travel a long distance from where you first hit the catching feature. Stop and now pace back to where you were. Think about the terrain you are travelling over to see if there are any clues you have missed.

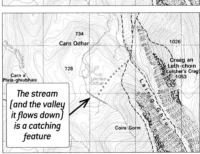

The stream (and the valley it flows down) is a catching feature

6 Pace the same distance in the opposite direction and look for a feature to take a fix from (see p. 125).

↘ EXPERT TIPS

➜ Before embarking on each leg, study the map for catching features you will expect to see along the way.

➜ A catching feature does not have to be a physical object – after a predetermined time/distance you stop and relocate.

TECHNIQUES

THUMBING THE MAP

A very simple technique to keep track of exactly where you are while moving without having to constantly be looking at the map.

Even though it is a very simple technique it is often overlooked; I use it all the time.

- Fold your map so it is more manageable.
- Find your exact location on the map.
- Hold your map in one hand and your thumb next to this location.
- As you move and identify features along the route, move your thumb next to these on your map to mark your new location.
- Keep your thumb on your map as a 'You are here' marker.

↘ EXPERT TIPS

→ This is a technique used by most orienteers who travel quickly across the terrain.

→ Some orienteers mark their thumbnail to create a more precise marker, either with a waterproof pen or sometimes even cut a small notch into their nail. I prefer the pen method!

MEASURING MAP DISTANCE IN THE FIELD

You generally need to calculate the length of each leg and the distance of the entire journey.

In planning your route you will already have calculated much of this information (for detailed methods refer to **Route Planning**, pp. 149–53). Once in the field, you may actually use different navigational legs and might even need to revise your route. The following techniques for measuring distance are the most practical in the field.

Using the compass roamer scales

This, in addition to **Visually Estimating Distance** (see pp. 168–71), is the method that I use the most. Your compass has roamer scale lines, with graduations marked to different scales – 1:50 000, 1:25 000 and 1:15 000.

1 Use the roamer scale corresponding to the scale of your map.

2 Place the compass over your map. Move the compass so that the zero on the scale is directly over where you wish to measure from.

3 Read along the scale to your destination and take note of the distance measured.

Using the compass ruler

If you are using a map with a scale that is different to those on your compass roamer scale you can use the compass ruler instead to measure distance. With metric mapping, dividing the map scale by 1,000 will give you the real distance in metres represented by 1 mm on your map.

Map Scale	1mm on your map represents	Map Scale	1mm on your map represents
1:10 000	10 m	1:25 000	25 m
1:12 000	12 m	1:30 000	30 m
1:15 000	15 m	1:40 000	40 m
1:20 000	20 m	1:50 000	50 m

Other techniques that are less practical to use in the field are detailed in the **Route Planning** section.

In hill and mountain areas you will need to take account of the additional distance covered travelling up and down a slope – the **Foreshortening Effect**.

TECHNIQUES

Calculating distance where terrain is subject to changes in elevation (foreshortening effect)

Slope can be given in two different ways:
- **Slope Angle** is the angle between a horizontal plane and the surface of the hill
- **Percent Slope** is how many metres you rise in every 100 metres.

The most straight forward way to calculate the extra distance you need to travel, whilst operating in the field, is to use slope angle. (To use percent slope, see page 152.)

Using the map

1 Choose the area for which you want to calculate the slope angle; the slope direction must not change by crossing the top of a hill or the bottom of a valley.

2 Measure the map distance you wish to travel and make a note of it.

3 Measure the distance between the contours and estimate the average contour spacing – if the slope is steep, do this using index contours (every fifth thicker line).

4 Refer to the table below and select the angle nearest to the one you have calculated.

5 Multiply the map distance you noted by the extra distance relating to this angle.

Slope angle	1:25 000 Single contour	1:25 000 Index contour	1:50 000 Single contour	1:50 000 Index contour	Extra distance
45°	0.40	2.00	0.20	1.00	**41.4%**
41°	0.46	2.28	0.23	1.14	**32.5%**
37°	0.53	2.66	0.27	1.33	**25.2%**
32°	0.64	3.20	0.32	1.60	**17.9%**
27°	0.80	4.00	0.4	2.00	**12.2%**
20°	1.07	5.34	0.53	2.67	**6.4%**
14°	1.60	8.00	0.80	4.00	**3.0%**
7°	3.20	16.00	1.60	8.00	**0.8%**

↘ EXPERT TIPS

→ Map distance tools have progressed in recent years but they still have limitations if maps are not perfectly flat and without creases, so use with caution.

→ The reality is that, with experience, you build up an internal database of interpreting distance by reading the map and combine this with two simple yet very powerful navigational techniques: pacing and timing.

PACING

On level ground we each walk a natural number of steps in a given distance.

In land navigation we determine how many paces it takes for you personally to walk 100 m, this is called your **Pacing Count** ... or your **PC**. This known number of steps can be used to determine how far you have travelled. This system of navigation is called **Dead Reckoning** and is one of the most important techniques you will learn. It is often used in poor weather/visibility and at night.

Determining your PC

Select a level piece of ground, a football pitch is ideal. Use your map to measure 100 m or better still, measure it out with your distance lanyard.

1 From your starting line walk at your normal pace and count your double step – if you start by putting your left leg forward first, count every step thereafter of your right foot only – this is a double step and counts as one pace.

2 Stop at the end of the distance and record this number.

3 Repeat this, as you need to relax into your stride, and make sure that you are both counting and pacing consistently.

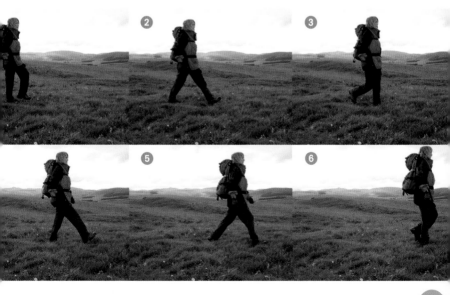

This number of paces is your personal PC (typically this varies from 55, for very tall people, to 75, for people with short legs).

You now know your personal PC

Calculating distance travelled ≤ 100 m
Now you know your own PC on this terrain, you can predict how many paces you will need to cover say 50 m (PC divided by 2) or 25 m (PC divided by 4).

Calculating distance travelled ≥ 100 m
When travelling further than 100 m it is easy to lose count so the technique employed is to count 100 m then, without stopping, start recounting your paces from one. Record each 100 m travelled on your tally counter.

↘ EXPERT TIPS

→ If you do not have a tally counter use small stones/pebbles. For 500 m pick up five and every time you reach your PC drop one stone and start to count from zero again. When you have dropped your last stone you have travelled 500 m.

→ An alternative to pebbles is to make a tally counter – put ten toggles on a length of paracord, attach them to your rucksack and move one down every 100 m covered.

→ On different slopes you can check how your PC varies accurately by measuring distance first with a 50 m rope.

For example:

- My PC is 60.
- To walk 450 m would be 4.5 x my PC (4.5 x 60 = 270).
- I start walking and when I reach 60 paces I click my tally counter and immediately start counting from 1 to 60 again.
- When my tally counter reads 4 I know that I only have 30 more paces to walk.
- I average a maximum of ±4 paces to my prediction which is 3% or a maximum of 13 m out over such a large distance. When used in combination with timing it is very accurate.

A good exercise to hone and refine your PC is to walk on level ground for a considerable distance recording progress on your satnav, so it does not matter if there are twists and turns on your route.

The following factors ***increase*** your pace count – you must allow for them by making adjustments.

Slopes	Moving up a slope
Winds	In a headwind
Terrain	Crossing sand, gravel, mud, snow or heavy undergrowth
Elements	Falling snow, rain, or ice
Apparel	Excess clothing, a heavy rucksack and boots with poor traction
Visibility	Poor visibility either in bad weather or at night
Alertness	Mental and physical exhaustion

When these conditions prevail reduce the distance you travel between attack points to help maintain accuracy. With practice you will soon learn to correctly estimate your PC in different conditions.

TIMING

If you know your speed, you can easily work out how far you have travelled by timing how long you have been walking.

Timing requires less concentration than **Pacing** and it is more accurate over distances greater than 750 m.

However, used together they are a terrific way of calculating how far you have travelled: the combination is vital in poor visibility conditions.

Speed

Knowing how fast you are walking is the key and speeds can be generalised as a:

- gentle stroll 3 kph
- good walking pace 4 kph
- fast walking pace 5 kph.

Approximate speeds for different terrains	
5 kph	Level surface covered in grass
4 kph	Variable, rough surface
3 kph	Soft snow/strong headwind
2 kph	Deep snowdrift/severe headwind

Lots of factors affect your speed so do not overestimate how fast you are travelling. These include load carried, prevailing wind and weather conditions, visibility – people walk slower in the dark – conditions under foot and most importantly the freshness and fitness of the party; speed slows down towards the end of a journey.

If you own a satnav, use this in helping you learn to judge speed, but remember GNSS speed is only accurate on level ground.

Distance calculations

Carry a card with the relevant calculations on it – it is a lot easier than trying to remember and in addition memory tends to fail when we are under pressure, so by carrying a card you can be sure your calculations are correct.

Examples: If your next **Attack Point** is 100 m and you are walking at 4 kph, it will take you 1 minute and 30 seconds to reach it.

- If you combine pacing and timing to reach the same attack point, it will take you 1 minute and 30 seconds to reach it and 65 paces (If your PC is 65).

→ Buy a cheap weatherproof digital watch with a stopwatch function so you can accurately start and stop timing and take account of any time spent stationary

- If your next attack point is 800 m and you are walking at 5 kph, it will take you 9 minutes and 46 seconds (300 m takes 3' 46" + 500 m takes 6" = 9' 46") and 520 paces (8 x PC of 65).

This procedure uses **Dead Reckoning** and is very reliable, trouble-free and precise.

TIMING CARD	SPEED				
DISTANCE	2 kph	3 kph	4 kph	5 kph	6 kph
50 m	1' 30"	1'	0' 45"	0' 36"	0' 30"
100 m	3'	2'	1' 30"	1' 12"	1'
200 m	6'	4'	3'	2' 24"	2'
300 m	9'	6'	4' 30"	3' 46"	3'
400 m	12'	8'	6'	4' 48"	4'
500 m	15'	10'	7' 30"	6'	5'
1,000 m	30'	20'	15'	12'	10'
Going uphill add 1 minute for every 10 m contour line crossed up to a maximum of 3 contour lines in 100 m					
Going downhill subtract 20 seconds for every contour line up to a maximum of 3 contours in 100 m					

Time to add for other conditions	
Darkness	$1/2$ daytime speed
Very heavy load >20 kg	$1/2$ normal speed
Heavy load	Subtract 1 kph
Headwind	Subtract 1 kph or more if very strong

This chart is available as a waterproof card from micronavigation.com.

TRANSIT LINES

When features on the landscape align one behind the other, they create an imaginary line called a transit line.

Types of Transit Lines

- Linear features such as a wall, a stream, the edge of a forest.
- The alignment of two visible features such as a building and another prominent land feature like a trig point.
- Contour shapes from a ridge to a valley.

Transit is made

A

↘ EXPERT TIP

→ Be imaginative and creative in making transits. A large part of skilled navigation involves creating 3D images in your mind.

TECHNIQUES

Application of Transit Lines

Such a simple technique has many excellent functions and can be employed as:

- A **Collecting Feature** when you cross them.
- An imaginary line you can follow as a **Bearing**.
- To confirm arrival at your **Catching Point**.
- To use as a **Back Bearing** for a **Resection** when by a linear feature.

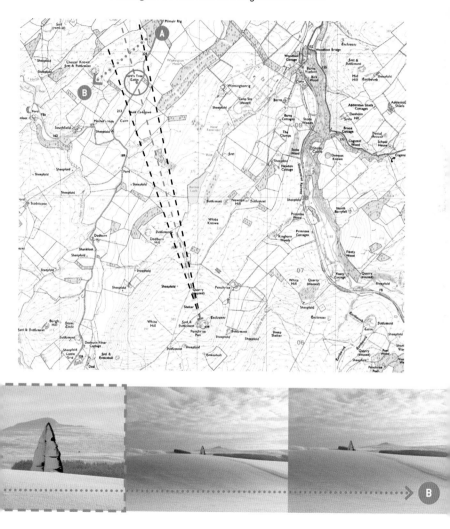

Good examples of where to use transit lines are:

- Walking off a ridge
- Following a bearing in an otherwise featureless environment
- Confirming you have arrived at a point of the map en route.

TAKING AND WORKING WITH BEARINGS

There are two ways of taking a bearing: one is by looking directly at the object and using a compass to walk to it, and the other is by using a map.

Taking a bearing to a visible object

Using a compass in the brace position

This is called a direct bearing and will enable you to hold your course even if the object is not always visible.

No matter which type of compass you choose to work with, whenever possible adopt the brace position to ensure you are as physically stable as possible.

Baseplate Compass

1 Assume the brace position facing the identifiable feature to which you wish to take a bearing. Place the compass on your knee and point the direction of travel arrow towards your identifiable feature. Let the compass needle float freely and it will point to magnetic north.

2 Move your head immediately over the compass housing to avoid creating parallax.

3 With one hand, hold the compass still on your knee and rotate the bezel until the red orienting arrow is exactly underneath the red north of needle. The north on your bezel will match the north of your needle.

4 Check again that the compass is pointing exactly towards your identifiable feature and that the arrow and needle are in perfect alignment.

5 The reading at the compass index is your magnetic bearing to this target. Do not move the bezel again.

Sighting Compass

1 Rotate the compass bezel until west (270°) is exactly against the compass index

2 Assume the brace position facing the identifiable feature to which you wish to take a bearing.

3 Raise the compass to your eye and point it straight at the feature – the direction of travel line will help you line up with it.

4 Look through the viewfinder and read the bearing. This is the bottom line of numbers, the back bearings are above.

5 Glance over the top of the compass to check that it is pointing exactly at your feature and then reread the bearing to confirm.

Mirror Compass

1 Open the mirror compass to 45° from the its baseplate. Assume the brace position facing the identifiable feature to which you wish to take a bearing.

2 Raise the compass to eye level and sight the feature through the sighting port at the bottom of the mirror – use the rear sight to align the compass up with it

3 Holding the compass steady in one hand, look in the mirror at the compass bezel and with your other hand rotate the bezel until the red orienting arrow is exactly underneath the red north of needle. The north on your bezel will match the north of your needle.

4 Check again that the compass is pointing exactly towards your identifiable feature and that the arrow and needle are in perfect alignment.

5 Lower the compass and looking directly at the baseplate take the reading at the compass index – just beneath the sighting port – this is your magnetic bearing to this target. Do not move the bezel again.

You can now walk on this bearing to your identifiable feature (see overleaf).

TECHNIQUES

Using a compass while standing

Sometimes it will not be practical to adopt the brace position and instead you will need to stand upright.

1 Stand face on looking at the identifiable feature to which you wish to take a bearing. Bring your compass to waist height and hold in both hands.

2 Point the direction of travel arrow towards your identifiable feature. Let the compass needle float freely and it will point to magnetic north.

3 Move your head immediately over the compass to avoid creating parallax.

4 With one hand hold the compass still and rotate the bezel until the red orienting arrow is exactly underneath the red north of needle. The North on your bezel will match the North of your needle.

5 Check again that the compass is pointing exactly towards your identifiable feature and that the arrow and needle are in perfect alignment. The reading at the compass index is your magnetic bearing to this target. Do not move the bezel again.

You can now follow this bearing to your identifiable feature.

Walking on a compass bearing

1 With the bearing set, hold the compass squarely out in front of you at about waist height and lean slightly over to look down on it. Let the compass needle float freely and it will point to magnetic north.

2 Rotate your body, not the compass, until the red end of the compass needle (North) is exactly over and aligned with the red arrow in the bottom of the compass housing. The front of the compass with the direction of travel arrow is now pointing towards your destination. Look directly in this direction and line up a distant landmark.

↘ GOLDEN RULES

→ Always estimate what you think the bearing is going to be before you take it with your compass.

→ If the bearing you then measure with the compass differs significantly from your estimation, question why this is before committing yourself to a navigation leg.

→ Learning to take bearings is the backbone of safe navigation and the more accurate you are the less error you introduce to the system of navigation.

3 Line up a distant landmark on your bearing and then, if possible, identify another nearer to you and which is exactly in line with the distant feature. Put your compass by your side and walk towards them.

4 Keep checking you can still see both objects and that they are in line. In this way you have created a **Transit Line** (an imaginary straight line) which will keep you on course.

5 If you cannot identify a distant feature, you can simply use one close to you but be careful of **Drift**.

Repeat the process to your next **Attack Point**.

Select features that are both clearly visible on the landscape, as well as on your map.

Working with bearings: using a map and compass.

Taking a bearing from an object on the map

1 Assume the brace position with your back facing the wind. Place the map on your knee. Put the compass on the map over the spot from which you wish to take a bearing, probably where you are (Point A).

2 Use the ruler line on the compass to join Point A to where you wish to travel (Point B) making sure that the arrow on the compass points in the direction you wish to go. Ignore the compass needle as it is not required for this technique.

3 Rotate the compass bezel until the N on the bezel points north on the map (always the top of the map). Align the compass housing orienting lines parallel with the map's vertical grid lines (north meridians, blue on OS).

4 The bearing to this object is indicated at the index, to follow this bearing on a compass you would need to adjust for **Magnetic Declination**.

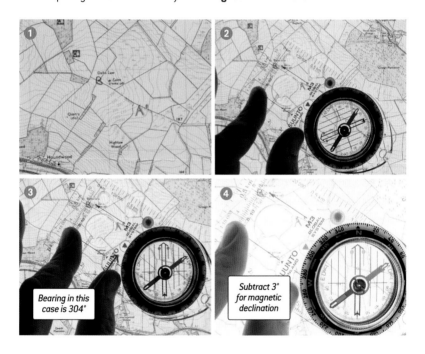

Bearing in this case is 304°

Subtract 3° for magnetic declination

Walking on a bearing taken from a map

If you wish to walk on a bearing, you will need to adjust for **Magnetic Declination** (see pp. 128–31).

1 Remove the compass from the map. Add the number of degrees declination by rotating your compass bezel if the declination is west; subtract the number of degrees if the declination is east.

2 You now have a bearing which has been transferred and corrected to use with your compass.

➘ EXPERT TIP

Transferring a bearing taken with your compass onto a map

1. Take a bearing to a visible landmark using the compass.

2. Adjust for magnetic declination.

3. Assume the brace position and place the map on your knee.

4. Identify where you are on the map and place the compass edge over this location (A).

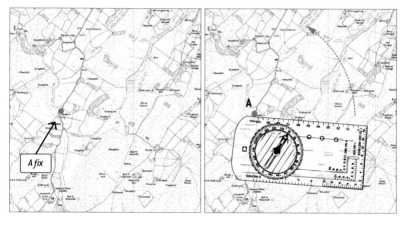

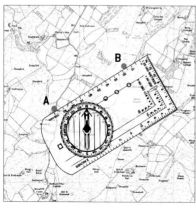

5. Using your location as a pivot, rotate the compass until north on the bezel is pointing to the top of the map and the compass housing orienting lines are parallel with the map's vertical grid lines. The edge of compass baseplate is now pointing in the direction of the visible landmark from which you took your bearing (B).

TECHNIQUES

BACK BEARINGS

Back bearings are invaluable – they are employed in leg confirmation, resections and in distance locator bearing.

A back bearing is in the opposite direction to your travel. So if you are following a bearing of 070° then to walk back along exactly the same route your bearing would be 250° – this is the back bearing (That is 070°+180°=250°).

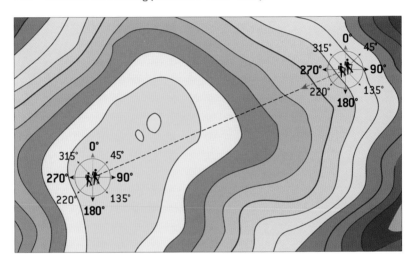

Leg confirmation

When you have arrived at your new **Attack Point** you can confirm that you have accurately followed the last leg. This may be necessary if your attack point is not a particularly obvious feature, such as a subtle contour change.

❶ Keeping the bearing set, hold the compass squarely out in front of you at about waist height and lean slightly over to look down on it. Let the compass needle float freely and it will point to magnetic north.

❷ Rotate your body, not the compass, until the WHITE end of the compass needle is exactly over and aligned with the red arrow in the bottom of the compass housing. This is your back bearing (180° to your original bearing).

❸ The Direction of Travel arrow should be pointing at the place you have just come from.

A quick way of executing this method is to turn and face the direction you believe your last attack point was and then see if the white needle of the compass aligns with the red north of the compass housing.

Resections

This is the use of back bearings to establish where you are. It is particularly effective if you are on or near a linear feature such as a path or the edge of a wooded area.

Quick method

1 From your linear feature identify a prominent feature on the land which will be on your map.

2 Face it and **orient your map**.

3 Orient your map and, looking down at the feature, follow an imaginary straight line back to the linear feature you are following.

4 Where your imaginary line intersects the linear feature is a rough guide as to where you are.

Precise method

1 Assume the **Brace Position**, facing an easily identifiable feature such as the summit of a prominent hill or the edge of a forest. Place the compass on your knee and aim it carefully at your identifiable feature.

2 Turn the compass bezel until the red arrow of the orienting lines is immediately under the red end of the needle (north). Now adjust the declination from magnetic north to grid north. Do not move the compass bezel again.

4 Put your compass on your map with the edge against the feature.

5 Using this feature as a pivot, rotate the compass until the north on the bezel is pointing to the top of the map and the compass housing orienting lines are parallel with the maps vertical grid lines.

6 Draw a line along the edge of the compass from the summit to your line feature, where they intersect gives you an estimated position (△).

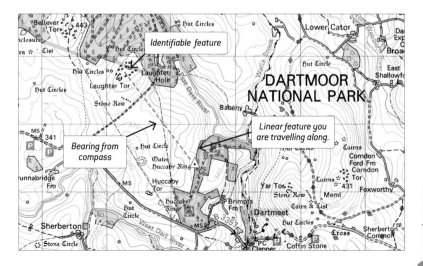

To obtain a fix (⊙), either identify another distant feature from which to take another bearing or with the information your EP provides, look for local features and employ **Terrain Association**.

Distance locator bearings

If you can see an individual or group of people clearly, for example if they were on top of a ridge, but they are unable to easily locate you because you are not silhouetted against the background, using either a compass, or ideally compass binoculars:

1 Take a bearing on the individual or party. Perform a **Stereoscopic Ranging** on them.

2 Communicate this bearing and the approximate distance either by radio or mobile phone to them.

3 They can now set their compass to this back bearing.

4 Using the compass as if they were going to follow a bearing, they rotate it in your direction and look straight ahead in search of you.

↘ EXPERT TIPS

The easiest way to think about back bearings is that whatever bearing the red north needle of your compass is pointing to, the white south needle (also sometimes black) will be pointing in the opposite direction – the back bearing. As you become more proficient a leg confirmation is rarely required. To calculate your back bearing:

1 If bearing is from 000° to 180° add 180° to calculate back bearing.

2 If bearing is from 181° to 359°, subtract 180°.

DRIFT

Every person has a natural lateral drift when walking. It is important to know what your drift is.

Everybody has a dominant side, in other words your right- or left-hand side is stronger, and for this reason when you walk you will put a little more power into the step on the dominant side, giving a slightly longer step. This causes people to walk in a curve away from the straight. Over large-enough distances (in a featureless desert) they will eventually travel a full circle. This technique can be best achieved in an area of level ground without obstacles – such as a large playing field or an open sandy beach.

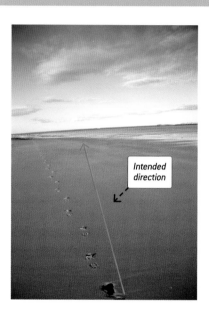

Intended direction

If you are by yourself, place an object to mark an edge, then walk from this point towards the centre of the area and double-pace the distance it takes to get there. It should ideally be at least 100 m (your pace count). Sight the object and face it, now close your eyes and pace 50 m towards it ... this will be unnerving!

Stop and check your drift from the centre of a line between your starting point and the object you were aiming for. Be aware of this; you always need to remember that you have a tendency to drift either right or left when following a bearing and take account of this.

↘ EXPERT TIPS

We are also subject to drift in the following circumstances:

→ On sloping ground: we naturally drift downhill.

→ A strong side wind: we will tend to travel off-course in the direction the wind is blowing.

→ Near areas of danger: when we are traversing an area where there is great danger to one side we naturally tend to drift away from it.

TECHNIQUES

ADJUSTING FOR MAGNETIC DECLINATION

The difference between magnetic north and true north at any particular location on the earth's surface is called the magnetic declination.

Magnetic declination is stated in the legends of most maps which use true north as the basis for the vertical grid lines. You correct for it simply by either adding or subtracting the number of degrees difference from your bearing.

If you are in the USA using a popular USGS 7.5' Topo map, the information is found near the bottom left corner and shows UTM Grid (GN), 1979 magnetic north (MN) in relation to true north.

If we took a bearing on a feature at, say, Silver Spruce with a compass and then transferred it to our map, but did not correct for the magnetic declination, it would be wrong by 13°.

To correct for it we need to add 13° from the compass bearing we took for working with it on the map. Likewise, if we took a bearing on our map and then wanted to use our compass to navigate to it, we would need to adjust for the magnetic declination – so in this example we would subtract 13°.

Grid North and True North

Lines of longitude run directly to true north. However, the National Grid lines are typically tilted a small angle from true north. The amount and direction of tilt depends on the location of that map with respect to the centre meridian of the zone.

On an OS map the only National Grid line running true north is the one that coincides with the longitude meridian 2°W, in the centre of this map projection. All of the others veer off either to the west or the east of this and the angle between these lines and the difference is called Convergence. In the UK True north differs from grid north by as much as 4°.

If you used the Magnetic Declination angle to correct your map and compass reading instead of the Grid Magnetic angle you have instantly introduced an error of more than 3° into your navigation. This is the first link in a potential chain of navigational errors, some of which are hard to correct for such as inaccuracy in taking a compass bearing, usually ± 3°, which when all combined will misdirect you at best and put you in serious danger at worst.

THIS IS VERY IMPORTANT
So for these maps we use something called the *Grid Magnetic Angle.* On OS maps, under the heading North Points (a grey shaded box) the Grid Magnetic Angle you need to adjust by will be described as:

> **At the centre of this sheet true north is 1°06' east of grid north. Magnetic north is estimated at 3°45' west of grid north for July 2002. Annual change is approximately 13' east.**

In 2002 we would have needed to subtract (*because the variance is west*) 3°45' from our compass reading to transfer it to the OS map. But if the current date is say 2010, we need to take account of the annual rate of change of 13' east. Thus 8 years x 13' change = 1°44' and because this change is east (*the opposite direction to west*) we subtract it from the 3°45' = 2°01'. So in 2010 we would use 2° for our corrections.

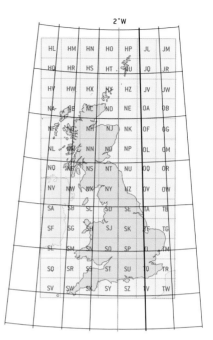

The British Grid meridians are blue; lines of latitude and longitude are black.

Presetting a compass for magnetic declination/grid magnetic angle.

Some compasses, such as the Suunto M3 which is used throughout this book, have an adjustable declination correction scale. This feature allows you to preset the compass to the local magnetic declination and eliminates the need to take account of the declination unless you move to an area where it is different. It is very convenient and if you are always navigating in the same region, as I am with my Mountain Rescue Team, then presetting the compass reduces the chances of making an error.

1 Check the magnetic declination in the area you will be navigating.

2 Rotate your compass bezel so the N is exactly aligned with the index triangle.

3 Turn your compass over and turn the small brass screw on the underside of the compass housing until the fine black line is set to this declination east or west.

Anti-clockwise: rotates the orienting lines west.
Clockwise: rotates the orienting lines east.

↘ THE RULES

If magnetic declination/grid magnetic angle is west

1 When transferring a bearing taken with your compass to a map, **subtract** the magnetic declination from your compass bearing.

2 When transferring a bearing taken on your map to your compass it is simply the reverse of the above – **add** the magnetic declination to your compass bearing.

If magnetic declination/grid magnetic angle is east

1 When transferring a bearing taken with your compass to a map **add** the Magnetic Declination to your compass bearing.

2 When transferring a bearing taken on your map to your compass it is simply the reverse of the above – **subtract** the magnetic declination from your compass bearing.

4 Turn the compass back over to its normal position and lay it on a white background – a sheet of paper is ideal. Very carefully check that it is the correct number of degrees (they are usually in 2° increments) and that you have also set it to east or west as specified on your map, website or from your manual determination of the difference.

There is a drawback which you must be aware of when doing this. It is very easy to stop being conscious of the effects of magnetic declination, so that when you move to another area you are liable to forget to reset your compass!

↘ EXPERT TIPS

→ If you are making a long trip double-check each new map's declination.

→ Only true north is correctly aligned in your map's legend; the others are not drawn to the correct angle – they are just a representation with the actual angle stated on them.

→ An easy to remember mnemonic for declination is: '*Declination east, compass reads least. Declination west, compass reads best.*'

→ Magnetic declination is also known as magnetic variation.

→ On OS Landranger maps (1:50 000), the difference between true north and grid north is given for each corner of the map.

TECHNIQUES

Finding out your magnetic declination

If the magnetic declination for an area is not marked on your map, there are a few ways to ascertain what adjustment needs to be made.

The internet
Global sites for magnetic declination include:

- **www.magnetic-declination.com**
- The National Geophysical Data Centre: **www.ngdc.noaa.gov/geomagmodels/Declination.jsp**
- Canadian Government's Earth Science Sector, Natural Resources Canada : **geomag.nrcan.gc.ca/apps/mdcal-eng.php**
- UK site for grid magnetic angle: **www.micronavigation**.

Your map and compass

1. Locate exactly where you are on the map; the more accurate the fix the more accurate your result.

2. Take a bearing with your compass on a distant feature, ideally one which is narrow such as a radio mast and which is identifiable on your map.

3. Make a note of this bearing.

4. Now take a bearing on the map from your position to the feature (on the map).

5. Compare the difference – this is your local difference to be applied to your compass/map. It will not be as precise as that stated on the map or internet but it will be adequate.

Your digital mapping or satnav
Some digital mapping programs and satnavs state the difference.

> ✖ **TIP:** You can also determine declination using the North Star/Polaris – See *Night-time Celestial Navigation* (p. 81).

BASELINES

A baseline is a bearing taken on a very prominent feature, such as a dominating mountain peak, that you can use to return to your start position.

Finding position using a baseline

Anywhere along this baseline you can take a second bearing from another feature and place yourself on the baseline.

1 Find the prominent feature you identified, even if you have to (safely) ascend to do so. If you are not exactly on this bearing, which is most likely, move sideways until you are.

2 If your bearing reads fewer degrees than your baseline bearing move left; if it is more move right.

3 Now take a second bearing of any other feature that will be on your map as near perpendicular to it as possible.

Plot these two bearings on you map. You now have a fix.

Be imaginative finding baselines, in urban navigation if you choose a skyscraper make sure it is distinctive enough from others around it. In areas where contours change little the baseline does not have to be particularly high, for example this water tower in northern France, which was clearly visible 20 km away.

Hook and baseline

When a baseline is combined with a **Catching Feature** this procedure is called a hook and baseline. I use it as a matter of standard navigational practice at the beginning of all of my journeys, because subsequently if I can see the prominent feature, I can align myself back onto my bearing on it and get back to my start position. And if, for whatever reason, I overshoot my start position, then the catching feature will alert me of this.

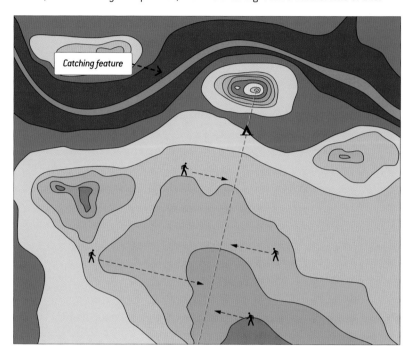

Catching feature

❶ From your start position, take a bearing on a prominent feature, one that you should be able to see at all points on your journey. This bearing is your baseline. Write it down.

❷ Study the map and select a catching feature that is behind the prominent feature from where you are.

❸ On your journey, if you need to make a direct route back to your start, or you are lost, locate the prominent feature you identified, even if you have to (safely) ascend to do so.

❹ Take your bearing to it. If your bearing reads fewer degrees than your recorded baseline, move left; if it is more move right.

❺ When your current bearing to the feature matches your baseline, travel towards it. On your journey back, if the prominent feature disappears from sight, continue to follow the bearing using **Attack Points** as you would normally use micronavigating.

Beware of hazards; it is rare that you can travel in direct line over a large distance on land so if you encounter obstacles/danger use **Boxing** techniques. You will eventually arrive at your start position.

The streambed, in the foreground of this view, could be used as a catching feature.

If you overshoot your start, your catching feature will stop you

1. When you reach your catching feature stop.

2. Calculate the **Back Bearing** of your prominent feature.

3. Now travel along your catching feature until you are on this bearing.

4. Travel back towards to your start.

↘ EXPERT TIPS

→ Employing this technique allows you to roam without a specific route, even off your map.

→ If you really are lost and unable to find your prominent feature start your **Relocation Procedure** and try to think of what obstacles may be blocking your view.

TECHNIQUES

READING A GRID REFERENCE

Grid references define specific locations. By knowing a grid reference you can locate a place on a map and travel to or from it, accurately describe it to others, possibly for a rendezvous or to let the emergency services know exactly where you are if you are in trouble.

In this section, you are going to learn how to read a grid reference from any metric map, including all European topographic maps, North American Topo maps and also how to read latitude and longitude coordinates – this should set you up to navigate anywhere in the world!

Knowing how to read a grid reference on a map is fundamental to all navigation.

They appear as a square grid, with grid lines numbered sequentially from the origin at the bottom left of the map (hence the term grid reference) and are numbered to provide a unique reference to an **area on a map**.

It is very important to remember that a grid reference places you inside a grid square (an area) – this is not an absolute position (pinpoint). The longer the grid reference the smaller this square. On any metric 1:25 000 scale map these areas are:

Grid Reference	Example	Using	Boundary of area covered (m)	Area described by the grid reference (m²)
4 figure	NT 54 16	finger	1,000 x 1,000	1,000,000
6 figure	NT 540 162	compass roamer	100 x 100	10,000
8 figure	NT 5405 1623	grid reference tool	10 x 10	100
10 figure	NT 54058 16239	GNSS/GPS	1 x 1	1

Giving the 'area described by the grid reference' a real-life meaning!

- A six-figure grid reference places you in an area larger than a Wembley football pitch.
- An eight-figure grid reference puts you inside an area the size of my kitchen!

Therefore, always obtain the most accurate grid references possible. Compass roamers will at best give you a six-figure grid reference (Wembley football pitch!) and it is easy to make mistakes with them, which is why I never use them.

TECHNIQUES

Instead I use a grid reference tool to take out the guess work and accurately read, plot and give grid references with confidence.

These tools can be used on any 1:50 000 or 1:25 000 metric grid reference system, *because all metric map grid systems are the same!*

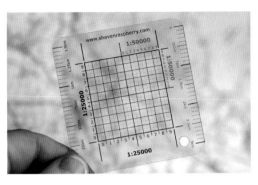

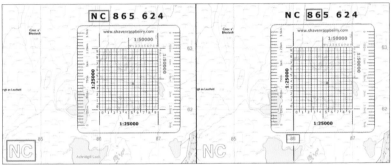

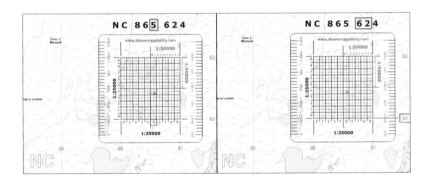

You now have a six-figure grid reference which you could also read (not as accurately) with your compass roamer. To make it an eight-figure grid reference, use the faint grey line as marking 5/10ths both easting and northing.

Now simply estimate how many tenths your easting is, then your northing. Your eight-figure reference (kitchen-sized area) is: NC 8567 6243.

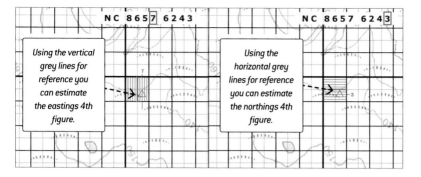

An easy phrase to help remember which reading to take first is: walk **into** the house and go **up** the stairs (which is you easting across the map, then your northing up the map).

Universal Transverse Mercator (UTM)

Taking a grid reference using the UTM is almost exactly the same as it is on a UK OS 1:25 000 – we deal with the individual grid squares in the same way. In **Mapping Systems** (pp. 39–43) you learned that the UTM grid system is divided into 60 zones. Each 'zone' runs longitudinally (north to south) covering 6° of the globe. Ten of these zones of these cover the USA.

The longitudinal lines dividing the zones are called 'easting'. The easting in the middle of the zone is called the meridian, numbered at 500,000 m. To the east of the meridian the easting will be greater than 500,000 m, to the west less than 500,000 m.

The latitudinal lines are the 'northing.' The middle northing line is numbered zero metres and is called the 'equator'. Northing lines north of the equator will be positive, and south of the equator will be negative.

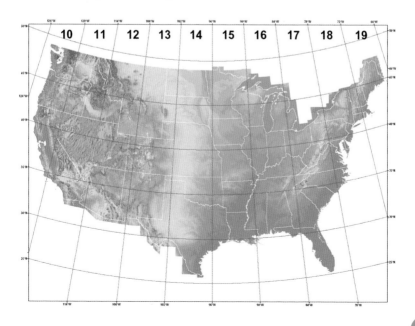

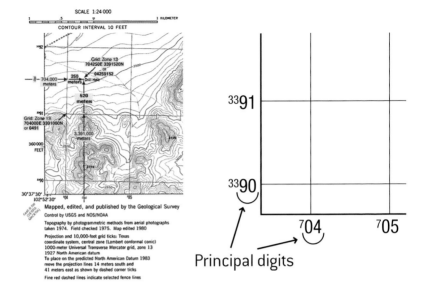

1 In the bottom left of the map read the zone you are in – zone 13 in this case.

2 Find the location you wish to take a grid reference for, in this case a drill Hole.

3 Read right to the grid intersection before your place of interest. In this case, it's line 704 – also known by its principal digits as line 04.

4 Count grid lines up to the intersection (for example, 3391 or 91). The abbreviated coordinate 0491 (think 04/91) gives the location to within 1,000 meters. Measuring right in meters from line 04, we find the drill hole is another 250 m. The complete easting component is 704250E.

5 Measuring up in meters from grid line 91, the drill hole is another 520 m. The complete northing component is 3391520N.

The complete grid coordinate for the drill hole is Zone 13 704250E 3391520N. This coordinate defines a location of the drill hole to within 10 meters.

Latitude and longitude

People often shy away from learning to use latitude and longitude as they think it is difficult in practice – actually, it's not and everyone should learn lat/long for the following reasons:

- every single spot on the earth can be identified by latitude and longitude
- it describes a precise point on the earth, not a grid square
- all air traffic use it
- all maritime vessels use it
- if you intend to travel to another country, you can use it
- many emergency services use it
- in SAR, all fixed-wing air assets use this system.

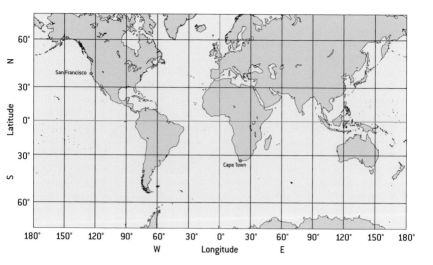

Firstly, a quick recap of what you learned in the **Global Mapping Systems** section.

Clearing up the confusion

Most people go wrong in determining latitude and longitude coordinates because it seems counterintuitive:

- latitude lines run east to west but are used to describe how far north or south a point is. The equator is 0° latitude and from here they go 90° north or 90° south
- longitude lines run north to south but are used to describe how far east or west a point is. The Prime Meridian (which runs through Greenwich, London) is 0° longitude and from here they go 180° east or 180° west.

So in the diagram overleaf the latitude and longitude of San Francisco, USA, would be 37° N 122° W, and Durban, South Africa, would be 34° S 18° E.

To make these readings more accurate, degrees can be subdivided in 60 minutes (') and in turn these minutes can be further subdivided into 60 seconds (").

Grasp these facts and you will have no problems in reading lat/long.

I always used to get confused because lines of latitude run across the world yet measure north to south – top to bottom – and lines of longitude run top to bottom yet measure east to west. The way I overcame this was to grasp the following simple detail:

- on each line of latitude all points are of an equal distance north or south
- on each line of longitude all points are of an equal distance east or west.

My home is in Scotland, which is in the northern hemisphere, so the latitude is going to be N something, and it is west of London, so the longitude is going to be W something. Now we find my house on the map and read across the top the latitude nearest to it, remembering that it will be running right to left (check the globes again!) – 55° N. Then we read across the side the longitude, remembering that it will be running from the bottom to the top – 2° W.

So the lat/long for my house is: 55° N 2° W.

Just like a grid reference (this is a coordinate system) we need to be as accurate as possible – we can read that my house is 26' further north than 55° N and 42' further west than the 2° W. So more precisely we can describe the coordinates of my house as 55°26' N 2°43' W

And to be really precise if we estimate how many 60ths further it is in seconds – we can precisely state that the lat/long coordinates for my house are 55°26'14" N 2°42'59" W ... and that is all there is to it!

Lat/long reading on maps based on grid systems

You will find that the vast majority of maps which use a grid system also have lat/long coordinated on the margins (edge) of the map thus you can also calculate a position using lat/long.

However, it is important to remember that this lat/long position will not be as accurate as your grid reference because, unlike the grid lines on your map which are absolutely straight, lines of longitude would curve if drawn on the map.

↘ EXPERT TIPS

→ Latitude is always given before longitude.

→ Degrees of N, E, S and W are sometimes referred to as negative degrees: San Francisco would change from 30° N 30° W to Lat 30° Long –120° and Durban from 30° S 30° E to Lat 30° Long –30°.

→ Latitudes are parallel, but longitudes are not.

→ Key longitude lines are: the Prime Meridian (0°) and the International Date Line (180°).

→ Key latitude lines include: the Arctic Circle (66° 33' N); Tropic of Cancer (23° 26' N); equator (0°); Tropic of Capricorn (23° 26' S); Antarctic Circle (66° 33' S).

BACK SNAPS

If you intend returning via the same route, at key points turn around and familiarise yourself with the view.

Make a mental note of what you will see at a point where your route changes direction; if it is an important junction make a note of it, including the grid reference, using your pencil and paper.

If your route is an expedition and expected to take days, weeks or even months, the best method is to take a digital photograph of the view and make a note of the photograph number next to its **Grid Position** on the map.

A digital camera makes a great addition to your navigation equipment.

TECHNIQUES

AIMING OFF

Aiming off is an essential technique to use when your attack point is not immediately visible.

In poor weather you can lose sight of even the largest attack point – small or subtle attack points (such as a contour change) can be missed even in the best of conditions. The technique involves aiming at a linear feature, deliberately to one side of the attack point, then simply following it to your attack point.

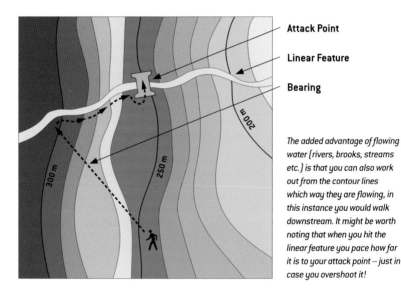

Attack Point

Linear Feature

Bearing

The added advantage of flowing water (rivers, brooks, streams etc.) is that you can also work out from the contour lines which way they are flowing, in this instance you would walk downstream. It might be worth noting that when you hit the linear feature you pace how far it is to your attack point – just in case you overshoot it!

Method

1. Select your **Attack Point**.
2. Identify a **Linear Feature** such as a wall, stream or track near to the attack point.
3. Take a **Bearing** to one side of your attack point.
4. Calculate from the map the approximate distance from the point at which you will reach the linear feature to your attack point.
5. Walk this bearing.
6. When you reach the linear feature use it as a **Handrail** to find your attack point, pacing the distance.

TECHNIQUES

BOXING

Moving from one attack point to another you may have to circumnavigate an obstacle.

To move around this obstacle and maintain your original bearing you use one of three box manoeuvres.

DO NOT MOVE THE BEZEL OF YOUR COMPASS FOR ANY OF THESE MANOEUVRES

Rough box

Used where you can clearly see to the other side of the obstacle.

1 As you approach the obstacle identify a very clear intermediate landmark that is on your course beyond the obstacle.

2 Stop at a point which you calculate you could look back at from the landmark on the other side.

3 Circumnavigate the obstacle to your new landmark. At this landmark turn around and confirm that you have reached it by taking a **Back Bearing** to your starting point.

Pure box

Used where you cannot see the other side of the obstacle

1 Stop at a safe distance when you reach the obstacle, facing it. Estimate the size of the obstacle either from the map – or visually if it is not on the map (e.g. rough ground or a bog).

2 The detour starts at right angles to the obstacle by choosing either east or west on your compass.

3 Rotate your body until the red north of the compass needle points to either east or west on the compass bezel. Pace this bearing until reaching the edge of the bottom of the obstacle and when it is safe to, walk forwards again and stop. Turn and follow your original bearing.

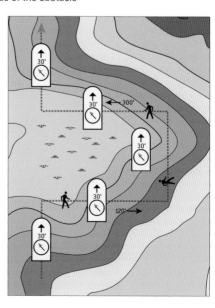

4 On reaching the edge of the obstacle, and when it is safe to walk across the top, stop. Then rotate your body until the red north of the compass needle points either east or west on the compass bezel – whichever is the opposite of the start of your detour.

5 Pace this bearing using exactly the distance that you originally paced and stop. Turn and follow your original bearing.

Stepped box

Used if the obstacle is particularly large and irregular in shape.

1 Stop at a safe distance when you reach the obstacle, facing it. The detour starts at right angles to the obstacle by choosing either east or west.

2 Rotate your body until the red north of the compass needle points to either east or west on the compass bezel. Counting your paces, walk in this direction until reaching an area where it is safe to walk forward again, then turn and follow your original bearing until reaching another area where you need to detour, and stop.

3 Again rotate your body 90° in the same direction as Step 2, and pace this bearing until reaching an area where it is safe to walk forwards again, stop and add these paces to your first east/west part of the detour.

4 Repeat this as much as is required to reach the top edge of the obstacle and when it is safe to walk across the top and stop.

5 Rotate your body until the red north of the compass needle to points either east or west on the compass bezel – whichever is the opposite of the start of your detour. Pace this bearing using the total east/west distance you covered and stop. Turn and follow your original bearing.

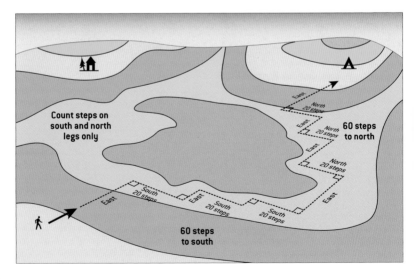

While on course, you run into a hill.

1 You take a 90° left turn and pace count until you clear the hill.

❷ Then turn right 90º and walk till you clear the hill again.

❸ Then turn right 90º again and pace count the same amount as the first pace count. At the end of the pace count, turn left 90º and continue on your course bearing.

While on course, you run into a lake.

❶ On the other side, identify some object you can clearly see on the same course bearing.

❷ Walk the lake shore until you get to the same object and continue on your course bearing.

In winter, when boxing around sheer drops, take a very wide margin from the edge because of potential cornices.

RADIAL ARMS

This is a technique which gives you a quick reference to the cardinals of the compass and help you maintain a bearing without constant reference to your compass or satnav.

When there is no obvious feature to use as an **Attack Point**, continual awareness of which direction you are facing and travelling is essential for safe and efficient navigation.

> 'Sir Ranulph Fiennes first introduced me to this technique which he first used in solo circum-navigating both poles.'

Celestial radial arms

The most common objects to use as radial arms are the sun and the moon. These celestial objects move slowly through the sky – a maximum of 1° every 4 mins. Depending upon how accurately you need to navigate the terrain, this method can be used for up to 1 km legs.

- If you are moving at 5 kph you would travel 1 km in 12 mins, during which the sun or moon will have only moved 3° at most around the compass.
- Travelling at 3 kph you would cover 1 km in 20 mins – maximum movement of 5°.

As micronavigation utilises short legs, for short time periods, you can effectively ignore these small changes.

Method

When you have set your direction of travel and are facing this direction, still holding your compass estimate the approximate cardinal position of the sun/moon in relation to where you are standing and remember this.

1. Reach up, as if to grab the celestial object, being careful not to move your feet and remember this position. If the sun or moon is behind you, use your shadow as a radial arm reaching out in this direction.

2. Lower your arm and start your leg keeping the sun/moon in the same relation to your body (like an arm on your shoulder maintaining your position).

❸ At the beginning of every leg repeat this exercise.

This is a very valuable technique and should be constantly practised by all navigators. It is particularly useful for SAR dog handlers who are gridding an area.

   If the sky is clear and unlikely to change soon you can also use a principal star within a stellar constellation or planet which can be both easily identified and seen, such as Venus in the northern hemisphere.

Terrain radial arms

Sand dunes in the desert are a principal navigation aid used as radial arms (See **Special Environments: Desert Navigation**). Mountain ridges and ranges can often run continually in the same direction and are useful.

ROUTES

Your objective can be as varied as a hilltop with a magnificent view to the site of an accident for SAR teams.

Rarely can you move across the land as the crow flies, so your journey will involve various detours to accessible points along the way; these are called **Attack Points** and form your **Route**.

This manual essentially teaches you different techniques, using a map, compass and satnav to reach your objective via the attack points.

Routes can be:

- Meticulous – pre-planned, marking out every attack point on your map, calculating distance to travel, difficulty of terrain and available daylight, plus weather forecasts.
- General – understanding your start and objective, marking key attack points and areas of interest or danger, taking into account the type of terrain and weather forecast.
- Field-based – when you arrive and then decide where you want/need to go, you can then prepare a general route which is continually updated and revised on the way.

Route Card		
Date		
Estimated finish time		
Cut off time for safe return & to call the emergency services		
Start time		
Start/finish location inc grid reference (GR)		
Leader's name		
Number in party		
Contact mobile numbers	Name 1	
	Name 2	
	Name 3	
	Name 4	
	Name 5	
Next of Kin Telephone numbers	Work Home Mobile	

Location (GR)	Attack point	Bearing	Distance	Time for distance

Escape Route

Leave a copy of this with someone who can check when you are meant to return and will call the emergency services at the designated time.

This card is available as a free download from micronavigation.com.

ROUTE PLANNING

The single most important consideration before setting out — you must have a route.

The easiest way to plan a route is by using **Digital Mapping**, but this section describes the steps you need to take whether or not you own this type of software.

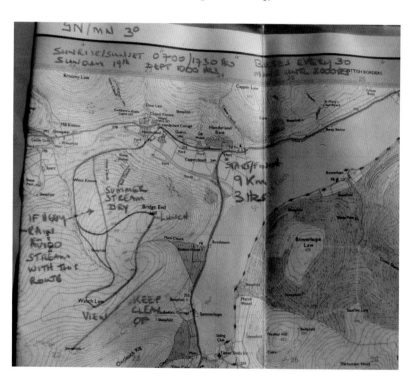

Route rules

- All routes should be flexible and changeable in response to factors which can vary unpredictably, from the health of individual members of the group to a change in the weather.
- A route should also create a story of the journey you are about to embark upon so that you have an image of what to expect both in terms of the land shape and features to the type of weather you will encounter.
- Plan to avoid errors: re-examine the route and try to determine where errors are most likely to occur and how to avoid them.

TECHNIQUES

- Every member of the group, whether an experienced navigator or not should be informed about the planned journey, how long it will take and what you expect to see on the way.
- Build in time for rests, breaks and sightseeing, and don't forget to take into account daylight hours and public transport times.

Familiarise yourself with the general features of the landscape. For example, mountain ranges tend to run in the same direction, rivers tend to run in a particular direction and cultivated land boundaries are often aligned with the cardinals of the compass. Check online for the prevailing wind direction. Make a note of these generalisations and add them to your **Environmental Clues**.

Route choice

Choosing which route to take involves the consideration of several factors. It anticipates potential obstacles and dangers and remedial measures to be taken, with safety being the single most important consideration in every journey.

- The journey time can be assessed against the available daylight hours and weather forecast.
- For SAR it is the safest means of getting to a location quickly.
- In leisure, it is will probably be for the views and the level of difficulty dependent upon the group's experience balanced against their aspirations.
- The choice of route has several factors from what equipment needs to be carried to the level of fitness of the group.

Route distance

Prior to embarking upon a trip, you should have a good idea of how far, over what terrain and in what conditions you will be travelling, taking full account of the weather forecast. This, factored in with the fitness of your group and the kit they will be carrying, will help determine both the feasibility and safety of your journey. Therefore, you should always allow for error by adding distance into your planning. You will need to calculate distance *before* your journey, *during* your journey and ideally *after* your journey, so that you can build up your knowledge of map interpretation related to land covered.

There are several methods for determining distance over relatively flat terrain and they are listed here, from the least to the most accurate.

- Using the grid lines on your map. These are equally spaced and by simply following your proposed route you can count the number the grid squares it crosses horizontally, vertically or diagonally to get an idea of the total distance.
- The ruler on your compass is a practical way of determining linear distance.

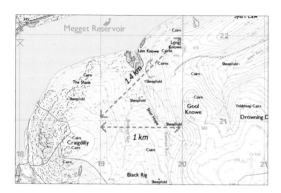

The distance across the diagonal on a 1:25 000 map (1 km grid) is 1.4 km.

- Use the edge of a piece of paper, by dividing the route up into small straight sections with a pencil then lay it horizontally over the map and count the number of grid squares it crosses.
- Take a piece of cord/string and lay it carefully over the track then check the distance by pulling the string taut and counting the number of 1 km grid squares it crosses.

↘ EXPERT TIPS

- Using a map wheel. This device uses a small rotating wheel that records the distance travelled. Distance is measured by placing the device wheel directly on the map and tracing the trail or planned route with the wheel.

Calculating distance where terrain is subject to changes in elevation

Travelling up and down steep slopes can add significant distance and this effect should be considered: it is called the foreshortening effect.

Calculating the foreshortening effect is often over-complicated – I only ever use the following simple method both on SAR missions and for leisure.

1. Choose the area where you want to calculate the percent slope; the slope direction must not change by crossing the top of a hill or the bottom of a valley.

2. Place the ruler of your compass perpendicular to the contours on the slope.

3. Measure the map distance from a contour at the bottom of the slope to the top.

4. Subtract the elevation of the lowest contour from the elevation of the highest contour .

5. Divide the elevation change by the distance of the line you drew. Multiplying the resulting number by 100 gives the percent slope.

6. Use the percent slope distance table beneath and multiply the measured distance on your map by the additional distance percentage.

Percent slope	Additional distance
20%	+ 2.0%
40%	+ 7.7%
60%	+ 16.6%
80%	+ 28.0%
100%	+ 41.4%
120%	+ 56.2%
140%	+ 72.0%
160%	+ 88.7%

A handy rule of thumb to remember – for every contour index line (the thicker brown line of 1:25 000 OS Explorer maps) add 5 mins to your journey time to the top or if there are many peaks add up all the ascents for the entire journey – think brown lines.

Escape Routes

Many factors can influence the need to get back either to your starting point or to safety quickly, such as the weather or injury. Marking potential escape routes prior to departing is immeasurably preferable to trying to calculate one in the heat of the moment. If you are traversing difficult or dangerous terrain, such as a mountain ridge, mark a few safe escape routes at different intervals along your proposed route.

Route Summary

A sample is available as a download at **micronavigation.com**. You should carry the details of the route you have created with you on your journey, printed on waterproof paper or stored in your map case.

Include on the summary either a sketch of your route with grid references, compass bearings and danger areas marked or a map overlaid with this information, plus:

- weather forecast
- sunrise/sunset times
- local magnetic declination (see **micronavigation.com**)
- telephone numbers of emergency contacts such as park rangers, or mountain rescue teams.

Personally, I like to print either the solar or lunar timetables onto the back of this summary if I plan to use **Radial Arms** and as a backup in case my compass fails. You could also add your **Pacing** and **Timing** chart.

Give a copy of your route summary to a trustworthy person and agree with them who they should contact if you do not arrive back at the stated time. Add the names of people in the party to the card.

Most digital mapping programs allow you to print out a route card, I then manually add any of the above information that is missing.

↘ GOLDEN RULES

→ Routes must always remain flexible to the individual's/group's requirements should factors change such as weather or the health of a group member.

→ Route cards can be left with a contact who can alert the authorities if the party does not return and give the authorities the plan.

→ Try to remember the shape of features rather than their colour as our brains retain shapes better than shades and hues.

→ Always allow for error by adding distance into your planning.

→ The hidden art to planning a route is using the map to visualise the terrain you will be crossing – the hills, valleys, paths you expect to see on your journey.

TECHNIQUES

3

SLOPE ASPECT

This is one of the most powerful navigational tools to both confirm your location, or, if you are lost, to find your position again.

You can use this technique anywhere and anytime where there are slopes and in any weather conditions, so long as you have visibility of at least 10 m. There are three stages to the technique.

Stage 1: Confirmation

When you merely wish to confirm your location en route.

1. Stop and face down the slope you are on.

2. Hold your compass away from your body and point in the direction of the fall-line down the slope. Note the nearest cardinal to this direction of travel.

3. Observe the steepness of the slope – if it is steep, expect to see the contours close together, and conversely if the slope is gentle more widely spaced. Set your map.

4. Search the slopes on the map in the area where you are for one which faces the same direction as the cardinal and where the contours approximately represent its steepness.

5. Look around you to identify other features, including terrain association, to confirm your **EP**.

Confirmation

Relocation

Stage 2: Relocation

If you are lost, use the following steps:

1 Face down the slope you are on. Hold your compass away from your body and point in the direction of the fall-line down the slope. Holding the compass level, let the needle float freely; it will point to magnetic north.

2 Holding the compass still, rotate the bezel until the red orienting arrow is exactly underneath the red north of needle: the North on your bezel will match the North of your needle. Check again that the compass is pointing exactly down the fall-line.

3 The reading at the compass index is your magnetic bearing. Adjust for magnetic declination. Do not move the bezel again.

4 Assume the brace position facing across the slope, back to the wind. Place the map on your knee. Put the compass on the map roughly in the area where you think you have been travelling.

5 Align the compass' red/black orienting lines north and parallel with the map's blue vertical grid lines. Move the compass slowly over the area, keeping the orienting lines parallel with the map's grid line. Where contour lines cross the edge of your compass at exactly 90° is the slope you are likely to be on.

6 To establish where you are on the slope, look at the ground around you for less obvious features, such as a slight depression or a small area of level terrain, and relate these to your map to confirm your estimated position.

7 Ideally, move along this contour, if safe to do so, until its aspect changes and repeat this technique to confirm your location. You can also determine the slope angle to confirm your location, see overleaf.

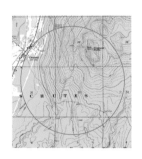

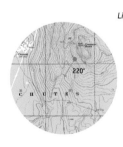

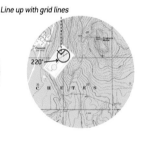

Line up with grid lines

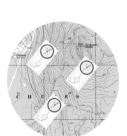

The baseplate is at right angles to the contours here

You have reduced your EP down to this circle

Stage 3: Determining elevation

Sometimes conditions such as a snow covering make it very difficult to identify small local features. By determining your height, you can convert an estimated position into a fix.

If you have an altimeter use it. If you are carrying a satnav you can confirm your height using its barometric altimeter. If you are relocating using slope aspect you probably have a poor GNSS signal, so don't use its satellite altimeter.

Alternatively, by measuring the slope angle you that are on and relating this to the contour spacing on your map, you can determine your elevation. To do this you can either use a compasses clinometer, such as on the Suunto M3, or by using your marked walking or ski poles, as described in the equipment section (p. 21).

Compass clinometer

❶ Rotate the compass bezel so that it reads exactly 270° (W) on the index triangle (where you normally read the bearing).

❷ Turn the compass onto its side and parallel to the slope, often best done at ground level.

❸ Read the angle of the slope on the back of the compass as indicated by the black pointer and make a note of this.

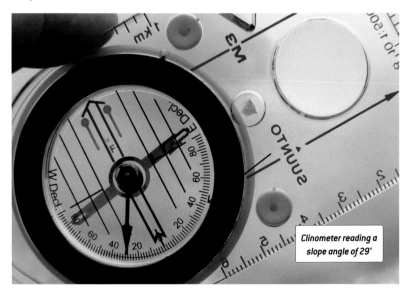

Clinometer reading a slope angle of 29°

Marked walking or ski poles (see p. 21 for marking up poles)

❶ Hold the marked pole vertically, handle up.

❷ Hold the other one horizontally at 90° to the first and slide the horizontal pole down until its tip touches the ground.

❸ Make a note of the angle on the vertical pole the horizontal pole sits.

From the table below, or using a slope angle tool*, confirm what the distance, in mm, between contour lines on the map should be for this slope angle.

Slope angle	1:25 000 Single contour	1:25 000 Index contour	1:50 000 Single contour	1:50 000 Index contour
45°	0.40	2.00	0.20	1.00
41°	0.46	2.28	0.23	1.14
37°	0.53	2.66	0.27	1.33
32°	0.64	3.20	0.32	1.60
27°	0.80	4.00	0.4	2.00
20°	1.07	5.34	0.53	2.67
14°	1.60	8.00	0.80	4.00
7°	3.20	16.00	1.60	8.00

Distance between contour lines on the map in mm where the contour interval is 10 m.

❹ Make a note of this distance. Refer back to the map and the contour line(s) that you have estimated you are on. Using your compass ruler or slope angle tool, measure the distance between the contour lines at these points on your map.

❺ Where the gap between the contour lines on your map matches the number you have noted, is your position (and therefore elevation) on this slope.

* Using a compass it is difficult to be precise with such small measurements. The author has developed a special tool to measure the distance between contours which is available from micronavigation.com.

↘ EXPERT TIPS

→ You can perform slope aspect up a slope as well as down. You may need to do this if you are at the bottom of the slope or have a restricted view below you.

→ To help visualise the direction down the slope, imagine you have just let go of a very large ball: which way would it roll down the slope? This is your bearing.

→ If you have an altimeter, read the stated height and search for a contour on the map near this height.

→ Sometimes this technique is referred to as 'Aspect of Slope'.

↘ WARNING

Parallel Errors (see pp. 185–6) are very easy to make using this technique, especially in poor visibility; therefore always walk to another nearby slope, where there is a definite change in bearing and repeat this technique or find another feature which is on the map.

DEAD RECKONING

This is the ability to move accurately from one point to another using only your direction and distance covered – in poor weather or low visibility it is an essential tool.

There are two types of dead reckoning:

- Forward dead reckoning – to predict in which direction and what distance you need to travel to reach **Attack Points**.
- Reverse dead reckoning – to retrace a route which you have just travelled.

The three fundamental stages of dead reckoning are:

1 Always start your leg from a confirmed **Fix** (⊙).

2 Accurately record distance and direction travelled on each leg of your route in a notebook.

3 Using this data, regularly update your position on the map.

LEG	FORWARD	BACK	Time
1	002°	182°	10 min
2	280°	100°	12 min
3	300°	120°	5 min
4	195°	015°	6 min

Sample of notes taken in the field, outlining, bearing, back bearing and time of each leg.

Forward dead reckoning

This is the combined use of **Pacing** and **Timing** on a specific bearing.

1 Identify exactly where you are – a fix. Identify on the map your next attack point.

2 Take a bearing from the map and measure with the compass roamer the distance to travel. Using your pacing and timing card (see p. 113), calculate the number of paces and how many minutes it will take.

3 Use your compass to establish your direction of travel.

4 Pinpoint a feature to focus on and move towards. Start your stopwatch, move off and begin counting your paces. If you pause remember to pause your stopwatch.

5 When you arrive at your attack point confirm your position using other map features.

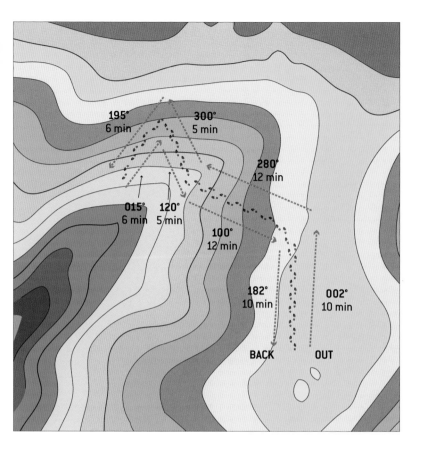

Reverse dead reckoning

1 From a fix ,start to record both the bearing and time taken to complete each leg.

2 Every time you change direction note the time taken to complete the last section and write down the new bearing; continue to repeat this process. Draw these on the map or write them down.

3 Travel at a constant speed. On completion of the route simply follow the **Back Bearing** of each leg for the time recorded.

Steep climbs/descents must be factored in where descending a slope will usually take less time than the ascent and *vice versa*.

Accuracy

The further you travel using dead reckoning the greater the chance for inaccuracy. Reaffirm your location after each leg or every 15 mins of travel.

RELOCATION PROCEDURE

Relocation is the procedure of using specific techniques to determine where you are if you are lost.

Lost is a relative word. You may not know exactly where you are on the map but have a rough idea of your location e.g. beside a river or even on a mountain you can name. Relocation from such zones is straightforward and the following techniques will enable you to do this.

Conversely, if you are totally lost and do not have even a rough idea of where you are, all is not hopeless and you can still probably relocate to safety or where you can obtain help.

Searching for this stream on the map to use as a linear feature.

The obvious

A psychological feature of being lost is to concentrate on the moment and the fact that what is immediately around you is not what you were expecting; your thinking becomes very localised. The most obvious and yet curiously sometimes overlooked technique is to identify any prominent distant features such as mountain tops and identify them on your map. You can perform a resection to determine your current location.

Actions

- As soon as you think you are lost **STOP** – do not venture any further.
- Take off your rucksack, sit down and have something like a chocolate bar and a drink, so you can collect your thoughts and relax.

- If visibility is poor due to driving rain, snow or fog, it may clear while you take stock and then you will be able to take a bearing again from a feature.
- If you are lost because of inclement weather, consider if you have the right kit and supplies to sit it out.

The following stages are listed in the order you should perform them. Most probably you will not need to go through them all; instead you will determine your location early on in the process:

- intelligence gathering
- static navigational techniques
- physical search
- equipment check
- calling for help
- moving off.

'This is the most important procedure you will learn and everybody gets lost sometimes. I have become lost in difficult areas to navigate, such as the jungles of Peru, to areas of hills around my home that I know well. The usual reason I have become lost is that I was not concentrating and this is the same for everybody."

Intelligence Gathering

- En route what did you pass? Think about any features, whether large or subtle, and **write these down**.
- Think about how long you have been travelling since your last known position and your approximate speed; **write this down**.
- Estimate how far you have travelled using the timing chart and **write this down**.
- If you are part of a group, ask them to help you with these first three steps **after** you have written down your own observations. Compare notes and discuss them.
- Remaining seated, identify on your map your last known position and draw a circle around it using as the radius the distance you have estimated that you have travelled.
- Study the map inside this circle because you are almost certainly somewhere within it.

Static Navigational techniques

- If you are on a slope and know which hill or mountain you are on perform a **Slope Aspect.**
- Set your map and keep your compass fixed against it.
- Holding your map and keeping it set, slowly walk around it through a full 360°, visually searching for features on the landscape while referring to your map.
- Try to identify a **Linear Feature** such as a ridge, road or track, contour feature, river or stream **near to you** and see how it aligns with the set map.
- If the visibility is poor, look for **Terrain Association** with the contours around you.
- In the vast majority of cases a reappraisal of the land around you and re-reading your map will give you at worst an estimated position and more probably a fix.

Physical Search

Using all of the above information, you need to consider physically moving from where you are to gather further facts. This must be controlled and only involve actions that can be retraced if they do not work. Again, follow these measures until you have determined your location.

TECHNIQUES

The most uncomplicated and dependable method is **The Wheel** (see pp. 181–2). I always use this first as it leaves me physically attached to the location where I first became aware of being lost and I am unlikely to compound the situation further by becoming more lost! If you do not have a rope or distance lanyard, never venture further than you can see from your position.

- If you are alone, mark your position with something like a stick or rock – do not leave your rucksack because *this is your life-saver!*
- Consider retracing your route back to your last known position, if you are sure that you can do this safely. When you arrive at this position, recalculate your navigation. On your return gather clues and stop and look for these on the map.
- If there is a linear feature near to you, such as a stream, rock face or track, walk over to it and perform a **Cliff Aspect** (pp. 174–5).
- Lastly undertake a **Spiral Box Search** (pp. 182–3), being sure to mark your starting position as above.

Equipment Check

There is a possibility that your navigational equipment may be faulty but it is more likely to be navigator error than defective equipment and, of course, it is highly unlikely that both your compass and satnav would be faulty at the same time. For this reason, this is one of the last things you should consider. If you only have a compass and think it is faulty, you should consider this stage as a priority.

Compass

Magnetic compasses can be influenced by many factors, so you need to eliminate these before you decide that it is not functioning correctly.

- When navigating under severe or difficult conditions, a simple yet frequent mistake is to forget that the north of your compass needle is red – check you are using the *red* end for north.
- If you are carrying a backup compass, check the readings are the same on both.
- Check for metal objects close by, such as a belt buckle, the under-wiring of a bra, or a nearby wire fence? If so, move it away from these objects.
- Make sure you are not underneath electricity transmission cables or within close proximity of an electricity sub-station.
- Have you moved into an area with large deposits of iron ore? Look at the geology of stones around you (most iron ores are dark reddish-brown) and on your map, because these areas are often marked.

Having eliminated these possibilities, put your compass in your pocket and set your map using :

- features in your surroundings
- stellar navigational techniques
- environmental clues

Keeping your map set, take out your compass and place it on the map and see if the red needle north matches your map's north (allowing for the local magnetic declination).

When you start to move again your compass may work once more if its deviation was due to environmental influences.

↘ EXPERT TIPS

→ People who create a plan and execute appropriate actions are the most likely to survive and the vast majority of people who go missing are found alive!

→ My MRT, along with most other SAR teams, find more than 95% of the people we are looking for. So call for help if you are in distress! We would always rather be called by you than weeks later have to recover people who did not survive.

Satnav

If you suspect your satnav is not functioning correctly, follow the following steps:

1 Recheck your satnav readings, holding the unit in the correct position away from your body. Statistically, human error is more likely than electronic error!

2 Give the unit as clear a view of the full sky as possible.

3 Check the battery level – many units lose accuracy when power is low and most batteries lose power in the cold. Replace the batteries if necessary.

4 Write down the reported grid reference, stated accuracy and height reading. Carefully mark this location on your map.

5 Using the compass ruler, mark the stated accuracy from this location plus 50%.

6 Draw a circle around the location using this mark as its radius.

7 Study the map details inside this circle, because in theory you are somewhere inside it, including its edges! If you think you have found your location, check that the height stated by the satnav matches the contour line height of where you are (± 2 times the stated horizontal accuracy) to confirm your location.

If after following these steps, you are sure that your satnav is inoperative and you are not carrying a spare receiver continue to follow the relocation procedures above.

Calling for help

Having exhausted all of these options, if you believe that you and your party are in a position of distress, do not delay and call for help using the **Emergency Calling Procedure**. SAR teams would rather be called before a situation becomes critical. In many cases they can often identify where you are from your description of the environment and guide you to safety on the phone.

Moving Off

If you are in a position of distress, are sure that you cannot weather the storm and are unable to contact the emergency services, you are going to have to move off. Think exceptionally carefully before embarking upon this last stage, as it carries the most risk and *never* move on a hunch, base your route on the following actions:

- Use all of your environmental clues to try and determine a route that will take you to civilisation and write them down. Spend time doing this, and then when complete, review either as a group or by yourself, if alone.
- In most countries most power-line pylons are numbered in sequence from the power station, a sub-station or near consumers of the electricity and start at No 1. Look at the spacing of the pylons and consider following them if the number is low. Most sub-stations have emergency phones if your mobile phone is not working.
- Find a water course and follow it downstream, because streams usually meet rivers and you are more likely to find civilisation near these. Be very careful it does not lead to very steep ground or waterfalls.
- Stop every time you encounter a distinctive feature; collect it and write it down. After a reasonable time, to allow your mind to settle, re-consult the map using these collected features.
- Keep your mobile phone turned off and try re-calling the emergency services every hour. Conserve your battery as much as possible.

LEAPFROGGING

Leapfrogging uses man-made markers to create Attack Points.

In bad weather/poor visibility this is a fundamental navigational tool, but it is often useful in good weather and visibility over large open expanses of terrain – from arctic environments to deserts, tidal deltas and sandflats.

Leapfrogging technique

Sending an individual ahead of the navigator when there are no other observable features maintains the correct bearing.

1. The person moving out in front needs to frequently look back for the stand still signal or, if they are not happy moving any further out, simply stand still.

2. The navigator then 'positions' them onto the set bearing. With this achieved the navigator walks towards the person out in front using them as an attack point.

3. If the same bearing is to be maintained, when the navigator reaches the person they continue past them, this time using the person as a back bearing (using the white end of the compass needle) to correct their course.

| Move this way | Move this way | Stand still | Come to me |

There are only four commands needed to position your leading person – any more overcomplicates the technique. It is vital that you agree these before separating. A good confirmation of understanding is to repeat the command to each other.

This technique can be repeated for as long as it is required.

Object markers

If navigating solo, create a small mound of snow or sand and either stick a twig into it or place a rock on top.

Move off and frequently look back and check the back bearing (using the white side of the compass needle) to maintain your course.

Key considerations

- The person sent forward must appreciate that, once in position, it is very important that they do not move until instructed.
- The greater the distance the more accurate the technique. In poor visibility, such as in mist or falling,snow, each step should be just short of the limit of visibility, or 50 m, whichever is the nearer. In a desert environment the step length is limited only by the ability to visually interpret arm signals – a maximum distance of about 400 m.

↘ EXPERT TIPS

→ This technique is often wrongly performed where the navigator, upon reaching the person sent forward, sends them forwards again – this halves the length of each accurate leg.

→ If conditions are severe attach a 50 m lanyard between each other.

→ When the forward person is near the limit of visibility, they should retreat back towards you a couple of metres in case the visibility deteriorates during the leg.

OUTRIGGERS

In poor visibility, having an understanding of changes in the local terrain is important.

With this technique, two people walk alongside you and near enough to be seen so you can gauge their height in relation to yourself – this will give you an appreciation of slope aspect and gradient.

Obviously, you need to be confident that there is no danger presented to your outriggers by features such as cliffs or bodies of water – if there is, rope up!

↘ WARNING

When travelling solo, it may seem tempting to use ski poles or other items of kit as object markers and move back and forth between them on your bearing: this is very time consuming and dangerous so should not be attempted.

TECHNIQUES

VISUALLY ESTIMATING DISTANCE

We can learn to estimate distance visually and the best way to practise is to take your map, stand at a known point and regard objects in your field of view, then measure on your map (using your compass' roamer) how far they are away from you.

Under certain conditions perceived distance can be deceptive. Objects will appear nearer to you if:

- it is a very clear day
- the object is brightly lit
- you are looking at them up or down a slope
- you are looking over water, snow or sand.

Distance	Clarity of objects viewed
50 m	Eyes and mouth of people can be clearly seen • Colour and type of clothing can be very easily identified • Hair colour can be determined.
100 m	Eyes can be seen as dots • Colour and type of clothing can be identified • Hands can be seen • General hair colour can be determined; hats can be seen if worn.
200 m	Faces and hands blur but recognisable • Colour of clothing can be identified • Number of people in a group can be determined • Rucksacks if worn can be identified.
300 m	No personal distinguishing features of people can be identified • Colour of clothing can be recognised • Number of people in a group difficult to accurately determine.
500 m	Livestock species can be easily identified, e.g. sheep and cattle • Overland power lines, the differences between pylons and poles easily identified • Differences in vegetation or crop can be recognised.
1 km	The ability to easily distinguish between deciduous and coniferous trees • Differences between dykes and fences can be seen • Livestock can be recognised • Trig points can be seen • Crags, and rocky outcrops recognised • Power lines recognised • Cars can be recognised.
5 km	Trees can be identified • Livestock can be seen • Houses and small buildings can be seen • Telephone masts and transmitters can be identified.
10 km	Large houses and towers can be recognised • Conspicuous hilltops can be identified • Telephone masts and transmitters can be recognised.

Available as a plastic-wallet-sized card from micronavigation.com.

And they will appear further away if:

- they are a similar colour to the background
- along a corridor such as a valley or even a straight road
- in poor light
- over undulating terrain.

100-metre memory

Visually judging distance is an important aspect of safe navigation.

1 Select an area of open parkland or a football pitch.

2 Get somebody to stand still and then move away 100 m.

3 Use a satnav to calculate this distance (the ground must be level as GNSS distance travelled is only accurate on level ground). Alternatively use a length of measured rope, a tape measure or markings on the football pitch and, if in parkland, features on your map.

4 Look at the person and write down exactly what you can and cannot see. This process will reinforce your memory.

This distance memory can now be employed to estimate distance up to approximately 800 m away by determining the number of 100 m increments to the distant object. With practice most people can learn to judge these distances ± 10%.

Stereoscopic ranging

This is a quick technique for estimating distance and can be accurate to ± 15%. Most people's arms are ten times longer than the distance between your eyes. Using this fact, you can estimate the distance between you and any object of approximate known size.

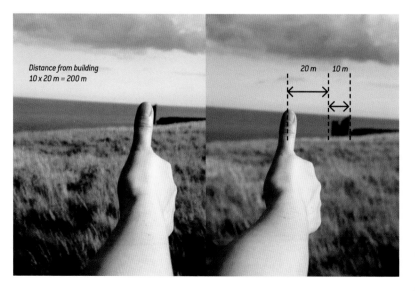

Distance from building
10 x 20 m = 200 m

20 m   10 m

1. Close one eye and look at your feature.
2. Raise your arm straight in front of you.
3. Hold up your thumb over the feature.
4. Close your eye and open the other one.
5. Estimate the distance your thumb appeared to move from the feature.
6. Multiply this by 10 to get a rough estimate of how far away you are.

Horizon estimation

1. Measure the height of your eyes, in meters or feet, from the ground.
2. Add your local elevation, in meters or feet, above sea level.
3. Multiply by 13 if you took the measurement in metres.
4. Multiply by 1.5 if you took the measurement in feet.
5. Take the square root to find the answer in kilometres and miles respectively.

This technique should only be used where the terrain between you and the horizon is relatively flat, such as a lake or desert.

Distance off

If you know a feature's height, or the width between two features, you can calculate your approximate distance using finger angles.

To estimate distance using a feature's height:

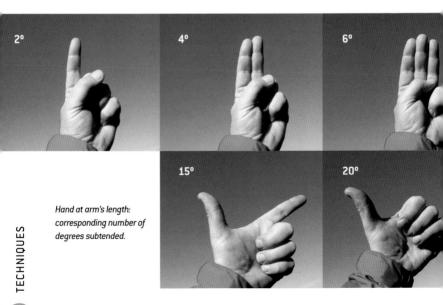

Hand at arm's length: corresponding number of degrees subtended.

Known height = 3,000 m
$$\frac{3,000\ m}{100 \times 6} = 5\ km\ distant$$

Known distance = 3,000 m
$$\frac{3,000\ m}{100 \times 15} = 2\ km\ distant$$

❶ Hold your hand out in front of you and rotate it through 90°.

❷ Estimate the angle from the top of the feature to the bottom.

❸ Divide the height by the angle and divide this by one hundred – this is your distance from the feature.

To estimate using width – simply rotate your hand.

↘ EXPERT TIP

→ For accurate calculation of distance see the section dedicated to using **Compass Binoculars** (see p. 63).

3

CONTOURING

Every time you gain height it takes more effort than moving on level ground; the steeper the ground the greater the effort.

In this example Route A is the shortest and because there are more than four contour lines in 100 m on the map it will involve some scrambling. So if you are a SAR team trying to reach a casualty at X quickly, some team members could take Route A while the stretcher carriers take Route B.

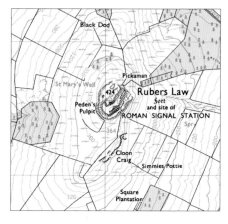

The choice of route depends primarily upon the level of proficiency over steep ground of the individual/group and how quickly you need to reach your target. So a hill walker would probably choose Route A.

Maintaining the same height (Route B) requires practice and it is important to establish a **Catching Feature** in case you go too far. A common way of navigating contours is to identify a distant object which is in line with the contour, use **Pacing** and **Timing** to reach the point when the line of the contour changes, then select another distant object and to keep repeating this exercise. Alternatively you can use **Transit Points** or **Collecting Features** if they are available.

An altimeter, or a satnav with an altimeter, should be used to check at regular intervals that you are maintaining height.

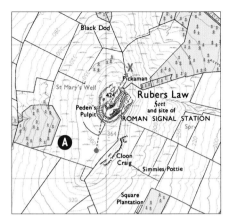

Route A: hard and fast

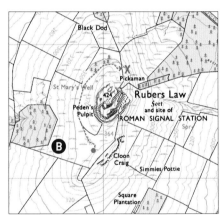

Route B: practice makes perfect

→ **Drift** is very common when first learning contouring – on your first attempts you will tend to lose height and then over next few attempts over-correct and gain height. However, it really isn't too difficult, it just takes practice and eventually can be undertaken in good visibility with no other checks.

→ The elite Russian Spetsnaz units have mastered this technique and use this along with very accurate interpretation of contours as their primary navigation tool.

→ A **Catching Feature** is a landmark which if reached defines that you have overshot your attack point. A **Collecting Feature** is a landmark which you pass en route to your attack point.

CLIFF ASPECT

This technique was developed for shoreline navigation although it can be applied to any linear feature inland.

Method 1 – Confirmation

When you are merely unsure of your position and wish to confirm your location:

1 Stop and face parallel to the cliff.

2 Note the nearest cardinal to this direction of travel. Set your map (orient north) – if the **Magnetic Declination** in your area is greater than 5° adjust the map's orientation accordingly. Search the cliffs in the area for one which runs parallel to your cardinal.

3 As you move continue to confirm your position using this technique.

Method 2 – Relocation

If you have landed on the shore by boat and do not know where you are, use the following steps:

1 Face parallel to the cliff. Hold your compass away from your body and point the direction of travel arrow directly along an imaginary line that runs parallel to the face of the cliff.

2 Let the compass needle float freely – it will point to magnetic north. Hold the compass still and rotate the bezel until the red orienting arrow is exactly underneath the red end of needle. The north on your bezel will match the north of your needle.

↘ EXPERT TIP

→ This technique can also be performed using linear features that change direction, such as rivers, roads and fences.

❸ Check again that the compass is pointing along an imaginary line which runs parallel to the cliff. Adjust for magnetic declination as necessary. The reading at the compass index is your magnetic bearing. Do not move the bezel again.

❹ Estimate how far the cliff runs parallel to your imaginary line using **Stereoscopic Ranging** (see pp. 169–70). Turn through 180° (the white end of your compass needle will now point to the bearing you have just taken). Estimate how far the cliff runs in this direct parallel to your imaginary line. You can now calculate where approximately you will be along this line.

❺ Assume the **Brace Position**. Place the map on your knee. Put the compass on the map roughly in the area where you think you are.

❻ Align the compass' red/black orienting lines northwards and parallel with the map's blue vertical grid lines. Move the compass slowly over the area keeping the orienting lines parallel. A cliff aspect that runs parallel to the ruler/edge of your compass is probably the cliff you are beside.

❼ Bear in mind your position along the cliff ascertained in step **❹**. Look at the ground around you for less obvious features, such as a wider area of sand or where sand becomes shale and relate these to your map to locate yourself. Mark this position on your map.

❽ If you are still unsure walk on until there is a definite change in bearing or another feature appears which will be on the map – identify this on your map and compare this with your other reading.

TECHNIQUES

BEARINGS ON THE MOVE

Occasionally there may be only one prominent feature mapped. You can still take bearings from this and estimate your location.

Running bearings

1. Take a back bearing from the prominent feature and plot it on the map.

2. Pace in a straight line until this bearing has changed by at least 30°.

3. Plot this second back bearing on the map. Record distance walked.

4. Using this distance on your compass ruler or roamer, move it from the feature parallel with the way you walked, until the distance on the ruler or roamer you walked fits exactly between the two lines.

The points of intersection are your locations when you took the two bearings. Use the last one as your EP (⚠).

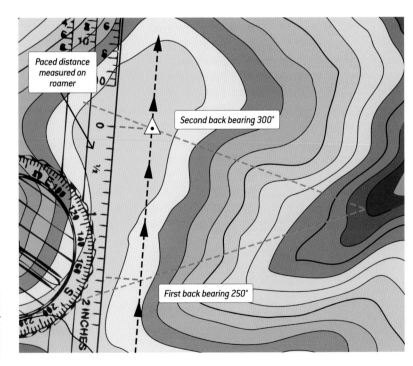

Paced distance measured on roamer

Second back bearing 300°

First back bearing 250°

Double bearing

This is a method for determining relatively accurately your distance from a feature and does not require plotting on the map.

1 Take a bearing on a prominent feature and note how many degrees it is different from your course – this is the relative bearing to the object and should be between 9° and 46°.

2 On the same course, start to pace count and keep taking bearings until your relative bearing is exactly double the first.

3 The distance you have travelled is the same now as the distance between you and the feature.

4 Draw this second bearing on your map as a back bearing from the feature and measure along it your known distance. This is your EP.

↘ EXPERT TIPS

→ If you can combine running bearings with a double bearing then you will be able to establish a **Fix**.

→ Bearings to distant objects can also be used when marking a prominent feature, which may be off the edge of your map when you are setting your **Hook and Baseline.**

Bearings to distant objects not on your map

In search and rescue you often need to take an accurate bearing on an object beyond the area your map covers. For example, another hill party you can see but who cannot locate you or, say, smoke from a plane crash.

Assisting a party to find you

Using only a compass – no map required

1 Ask the group to stay put.

2 Take a back bearing on them.

3 Convey this back bearing to the group. They can now either visually search for you looking in this direction or walk this direction to reach you.

This technique requires no map and does not require you to know where you are. At night time can be performed using head-torches or flares even if only one party is carrying these.

Using only the map – no compass required

1 Identify where you are on the map.

The smoke is originating in an area beyond your map. The golf house is exactly in line with the smoke and therefore on exactly the same bearing as the smoke.

2 Sight the distant object and while looking at it, identify a feature that will be on your map that lies in line between you and the distant object.

3 Plot a line from your location to the feature you sighted.

4 Estimate this bearing. You can use your wristwatch to help you do this by laying it beside the plotted bearing with 12 o'clock pointing to the top of your map; each hour represents 30°.

5 Convert this to a back bearing and relay to the other party.

Describing accurately the bearing of an object not on your map

1 Identify where you are on the map.

2 Sight the distant object and while looking at it, spot a feature exactly in your line of sight in-between you and the distant object that will be on your map.

3 Plot a line from your location to the feature you sighted.

4 Measure this bearing on the map using your compass.

5 Relay this bearing and your grid reference to the other party. They can now search along this line (bearing) for the incident locus.

Bearings to sounds

From time to time you may need to identify the location of a sound, for instance the six whistle blasts repeated every minute which is the standard signal used by mountain rescue. With practice, this method can be accurate to +/−10°.

1 Cup both hands behind your ears. Rotate your body, moving your feet and keeping your head fixed forwards, towards the sound.

2 Make small adjustments, moving your feet and turning your body towards where the sound is most intense.

3 Keep standing and take a bearing in the direction you are facing.

⬎ WARNING BEARINGS

When navigating in an area where there is danger (such as a ridge, cliff edge or a marsh) relatively close to either one or both sides of the bearing you intend following, you can create warning bearings.

1 On your map take a **Back Bearing** from your **Attack Point** avoiding the edge of the dangerous feature and move the compass ruler so it allows you a good margin of error.

2 Do this for the other side of your bearing, if dangerous ground lies this way too – note both of these bearings.

3 You need to travel to the attack point in-between these two bearings – move sideways to get onto this bearing.

4 Now follow this bearing and if your stray either higher or lower than your warning bearings, stop and move sideways back onto your original bearing.

5 Continually stop and check your bearing; this will minimise your **Drift**.

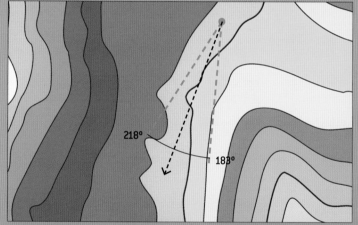

4

SEARCHING

Put simply this involves looking for something in a conscious and careful way. The most important aspect of any search is to be methodical and systematic.

Small or isolated features can be difficult to find and this situation only becomes worse in conditions of poor visibility. As you become a more competent navigator, you will start to select legs which provide a more direct route to your ultimate objective, where **Attack Points** can be anything from subtle changes in the contours to the type of terrain you travel across. Inevitably the situation will arise where you cannot immediately locate your attack point and will need to search for it. Six different techniques are outlined below, from the easiest to the most difficult. Choice of technique will depend upon such factors as whether you are in a group or not, or how poor the weather and visibility are. Master them all — one day your life could depend them.

1. Look again
Looking again is probably the most disregarded method of searching for your attack point.

- Stand still and this time slowly look around you through 360° searching for features and *using your environmental clues*.

In the vast majority of cases, a reappraisal of the land around you and re-reading your map will give you at worst an EP and more probably a fix.

2. Mapping
If at any point during your journey, you are unsure where you are, it is worth stopping to re-establish position, not doing so can compound the situation. This technique is also a very quick method of determining if you have arrived near to your **Attack Point**. If you are unsure:

1. Stop travelling.

2. Keep your compass set to the bearing you were travelling on.

3. Turn and face 90° to the left or right of your bearing. Pace this for a short distance until you have gathered more features, in particular look for contour changes and stop.

5. Now turn back and face 180° to the bearing you have just followed and pace the same number of steps back to where you have just come from.

6. If you have enough clues to help you relocate, continue again on your original bearing.

7. If you do not have enough clues, conduct the same exercise, this time going right if you went left or *vice versa*.

High Ground Search

A relatively easy and quick search to undertake if you do not need to gain a great deal of height – think about the effort of the climb and consider other search techniques.

1 Mark your position by placing a stick or something in the ground – never leave your rucksack.

2 Look for higher ground and take a bearing on this.

3 Pace to it and stop.

4 Repeat **Look Again** strategy.

5 If nothing, return to your mark and start your **Relocation Procedure**.

The Wheel

This is primarily employed either to confirm your position, converting an EP into a fix or, if you are lost, help establish your current location. With this technique you are effectively creating a mini map of an area of almost 8,000 m^2.

1 Secure the paracord to your rucksack and also to yourself. Using your compass cardinal north, move out in as straight a line as possible. Count the metres on the paracord as you travel.

2 Use your grease pencil and grid paper (or the back of your map) to record any change in slope angle and features, such as streams or waterfalls on your grid paper, also noting the distance away from your central point.

3 At 50 m return to your starting point, using the paracord to help direct you.

'When navigating in the dense jungles of the Amazonian rainforest in Peru, I developed a new search technique called The Wheel which I now instruct as the mainstay technique to solo navigators.'

TECHNIQUES

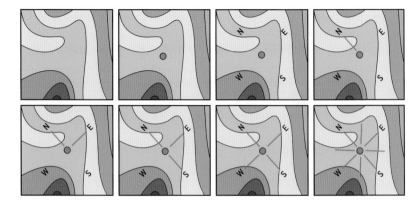

4. Now repeat this using the east, south and west cardinals of your compass, until you have collected enough information to help you arrive at a decision.

5. If you are still unsure, use the NE, SE, SW, NW cardinals to augment your map.

Spiral Box Search

This sounds more complicated than it actually is, so work from the photographs.

1. Mark your position by placing a stick or something in the ground – never leave your rucksack.

2. Pace out on any cardinal direction to within the limits of your visibility if the weather is poor and stop.

3. Turn 90° clockwise and pace the same distance and stop.

4 Turn 90° clockwise and pace double the distance and stop. Turn 90° clockwise and pace this new distance and stop.

5 Turn 90° clockwise and pace the original distance and stop. Turn 90° clockwise and pace the original distance and stop, you should now be back at your stick.

6 If you do not find your attack point start your **Relocation Procedure** (pp. 160–64).

Line Search (Sweep Search)

This is the traditional search with which most people in a group are familiar, yet it needs to be practised very carefully. A line of people walk, spread out at equal distances, a given distance in a straight line and then turn to walk back in the opposite direction, but this time covering adjacent ground.

1 Decide who is going to lead the search and how far back you wish to search. Study the map to identify potential dangerous features around you.

2 Form a line perpendicular to the bearing you arrived on, with the leader in its centre and on the spot you have arrived at. Space individuals regularly – well within view of each other.

3 Walk back the way you came, in a straight line and at the same steady pace, for the given distance, to see if you overshot your attack point and stop.

4 If the feature has not been found, all rotate through 180° and walk back double the distance on your original bearing, to see if you undershot your attack point.

5 If you still have not found your attack point, move back to the spot you started the search from and begin your **Relocation Procedure** (pp. 160–64).

It is very important to agree to a communication system prior to commencing and that the leader is in sole charge: for example, the line search leader says 'Hold the line' and everybody stops. The line search manager remains in the middle of the line controlling both sides of it and always moves at the slowest person's rate.

↘ EXPERT TIPS

→ Formal searches are time consuming, so try to relocate before you start one. Think about the terrain you have crossed, search on the map for small features which you would have passed to try and establish exactly where you are.

→ Do not cut corners with **The Wheel** search. When you reach the end of a specific spoke, do not try to walk back to your rucksack on a different spoke because:

- you see features differently on the return journey
- your rope may snag
- you will unnecessarily complicate the technique when you need to be thinking clearly.

PARALLEL ERRORS

This is not a technique! I have decided to include it in this section as it is the single most common navigation error that people make.

I have been on many MR missions where this emerges as the primary error that led to a chain of catastrophic events. Knowing how to avoid this error is very important.

Parallel errors occur when you mistake one feature for another – this can easily happen in poor visibility, or following a lapse in concentration. The difficulty is that you will continue to travel on the correct compass bearing and be fooled into making the map fit the environment. I did this on a ridge in Iceland when climbing with Sigurour Jonsson and Stefan Markusson, members of the Reykjavik HSSK MRT – they kept me on track!

1 If you suspect you may be running parallel, stop, check the grid reference on your satnav and transfer it to your map. This is the most frequently overlooked way to instantly check your location.

2 Confirm your height using either your satnav or an altimeter if you are carrying one. Then look at your map and find where you think you are and compare the nearest contour line to this with the device's reading – if it is more than ± 15 m difference, relocate.

3 If you are not carrying either of these devices set your map and firstly look at local micro-features. Examine the contours not just immediately in front of you but off to your sides and behind you; are there any sudden changes which would be clear on your map – a sudden drop, area of level ground that will confirm your location? Now look for distant features such as mountain tops; do they correspond with your set map?

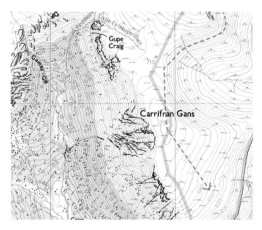

Planned route takes you up to the peak of Carrifan Gans and then downslope to the southeast.

TECHNIQUES

At the head of the valley, the planned change of bearing takes you towards the summit and from there to descent. But, if you overshoot before changing course ...

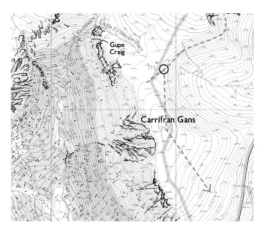

The green route is where you think you are as the land climbs, however, the red route is actually where you are – heading for a sheer drop!

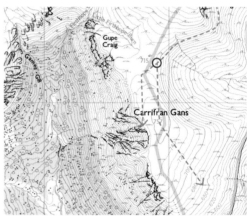

4 Create a mental list of features you would expect to see, including how the slope will change as you walk along it. Tick these off and if the next one you are expecting does not appear, stop and relocate.

> ✗ **HINT:** Other common errors are detailed in the **Appendix: Planning and Preparation** – if you don't know what to avoid ... you will probably encounter it!

WORKING WITH GRID REFERENCES

Be clear! If somebody gives you a grid reference verbally, either in person or via a phone/radio, repeat it back to them to confirm that you have heard it correctly and always give the two prefix letters, because the numbers alone could relate to any map within a mapping system.

If you are reading a grid reference from a map in order to put it into a satnav you will notice that your receiver asks for a ten-figure grid reference and you only have eight. Simply add a zero to the end of the easting and one to the end of the northing. You can now enter this grid reference into your satnav. **Example:** NC 8567 6343 becomes NC 85670 62430

Similarly if you take a grid reference from a satnav it will be ten figures yet you can only plot an eight-figure grid reference on your map. Simply round up or down the last digit of the easting and then the last digit in the northing. **Example:** NT 57631 20447 becomes NT5763 2045

If you wished to give the reference to somebody who can only use six-figure grid references (say with a compass roamer) the same reference would change from NT 57631 20445 to NT 576 204

Twelve-figure British grid references sound confusing – but they aren't. Some agencies in the UK, such as the police and fire departments use this system. It is exactly the same as the Ordnance Survey grid, only with the two prefix letters replaced by numbers.

So the NT box in the diagram becomes 3 and 6. So, for example, the OS grid reference NT 57631 20447, becomes 357631 620447.

↘ EXPERT TIP

→ There are 13-digit grid references that are required for locations in most of the Orkney Isles off the north of Scotland and north of there. The cathedral in the capital city of Orkney, Kirkwall is at: HY45786 15247 or 345786 1015247

CONVERTING GRID REFERENCES

There are many different regional grid reference systems in use around the world and converting between them is not practical to do manually, as doing so involves some extremely complex maths.

Luckily there are three very easy ways you can convert grid references which do not involve you racking your brains!

1. On your satnav you can simply change the **Position Format** to the reference system you wish to use – the device will now display locations in this mapping system. Many receivers will simultaneously display both lat/long and a position format of your choice.

2. Most paper maps have latitude and longitude marked along the edges of the map in addition to the regional grid reference system. OS Landranger and Explorer maps both have this information.

3. There are numerous websites which also allow you to convert grid references which you should be able to find using one of the leading internet search engines. An example in the UK is **www.streetmap.co.uk** where you can convert grid references into any of the following:

- OS Grid
- Landranger Grid
- lat/long
- military grid
- even a Post Code!

↘ EXPERT TIP

→ You may find an inaccuracy with different grid references for exactly the same location because they may be based on different datum and map projections – the British National Grid is based on a map projection only covering a small area of the globe, whereas longitude and latitude covers the entire globe. Whenever possible use the national grid reference system of the area you are navigating in.

SECTION THREE
**SPECIAL
ENVIRONMENTS**

INTRODUCTION

At first glance this seems a pretty daunting section, designed for intrepid explorers and professional SAR teams.

While I sincerely trust that these groups will find it valuable, the techniques used are exactly those which you have learned in this book. My *raison d'être* in writing this work has been to make passage on foot safer for anyone, anywhere and at anytime in the world.

Forests, do not have to be the Taiga Forest which has one-third of all the world's trees, nor do mountains have to be only the fourteen peaks exceeding 8,000 m. Some of my favourite walks are in the North York Moors of England and the skills from both the **Forest** and **Mountain** sections can be employed with equally good effect there as my visits to the Alps of Europe.

In each particular location I recommend the individual techniques which I have learned from the experts in these areas and later used when navigating solo through these regions. I do not repeat the exact procedure for each of these techniques, rather suggest why they are appropriate, unless there is a noteworthy variation of a technique which is better suited to the environment.

Preparation is the key: the right kit, the right tools, the right skills and the right mindset. Invest in the right equipment and learn how to use it in a safe environment first. There is no reason whatsoever why you should not practise **Pacing** and **Timing** in your local park and taking a bearing on your neighbourhood church, perfecting basic skills and starting small – I did! Do not attempt to scale some massive mountain until you have safely reached the summit of some less challenging ones and feel confident in your own ability.

For those people who do want to navigate through more challenging environments make the time to go on an introductory course on navigating in one of these regions. For example, in Scotland a day's introduction to mountain navigation in the company of an experienced mountaineer costs about the same as a night in a bed and breakfast; well worth the money.

Most of all enjoy the freedom that safe navigation brings you.

> ✗ **EXPERT FACT:** Some of the most extreme environments that I have visited have been comparatively easy to navigate: the difficulty was moving safely across them, mainly due to weather, exposed height or, in the case of jungles, creepy crawlies.

NAVIGATION IN EXTREME COLD ENVIRONMENTS

Alpine

An alpine climate describes the average weather (climate) for a region above the tree line. This climate is also referred to as mountain climate or highland climate.

The climate becomes colder at high elevations and this characteristic is described by the lapse rate of air: air gets colder as it rises, since it expands. The dry adiabatic lapse rate is 10° C per 1,000 m of elevation or altitude. This relationship is only approximate, however, since local factors such as proximity to oceans can drastically modify the climate. The main form of precipitation is often snow, often accompanied by stronger winds which also have a tendency to increase with altitude in mountainous regions.

The author at altitude: Bear Creek Spire, Sierra Nevada, California.

Polar

The two largest land areas with extreme winter conditions are the Arctic, the region around the North Pole, and the Antarctic, the region around the South Pole. However, during winter, many other areas of the world from Scotland to New Zealand can experience similar extreme winter conditions. The terrain in both regions may include plains, plateaus, hills or mountains and both regions experience massive variances in daily sunlight; this is a critical consideration in choosing when to navigate.

The Arctic Circle is one of the five major circles of latitude, and like the others is determined by the earth's relationship to the sun. North of this circle (latitude 66.56°N) the sun is above the horizon for 24 hours at the June solstice and below the horizon for 24 hours at the December solstice. It is this extraordinary change in the daily hours of sunlight which creates the dramatic weather conditions (see **Celestial Navigation**, pp. 69–71).

This area includes the Arctic Ocean, which overlies the North Pole, parts of Canada, Greenland, Alaska, Iceland, Norway, Sweden and Finland.

The Arctic region consists of a vast ice-covered ocean surrounded by treeless, frozen ground with broadly long cold winters and short cool summers. There is a large variability of climate across the region. All parts experience long periods of surface ice and some are covered with sea ice, glacial ice or snow year-round. Summer and winter conditions vary considerably with temperatures as low as –68° C and as high as 14° C.

Antarctica, the earth's southernmost continent, is almost entirely south of the Antarctic Circle (latitude 66.56° S) . Some 98% of the Antarctic landmass is covered by ice, which averages 1.6 km in thickness. It is the coldest, driest and windiest continent, with very little precipitation, except at the coasts; the interior of the continent is technically the largest desert in the world.

Temperatures vary greatly – from –85° C in the interior in winter to a maximum of 15° C near the coast in summer. Weather fronts rarely penetrate far into the continent, leaving the centre cold and dry. On the coastal portion of the continent heavy snowfalls are frequent – falls of over 120 cm in 48 hours have been recorded.

There are some advantages to navigating in extreme cold climates, particularly in the polar regions:

- low sun casting long shadows, highlighting contour features enabling **Terrain Association** techniques
- long day-length (can be 24 hours) and bright moonlight for night-time navigation
- hard underfoot with lakes, rivers, streams, swamps often frozen over and passable by foot, snowmobiles or sled
- unobstructed horizons facilitating excellent GNSS reception
- clear air allowing high visibility – giving long-distance views.

Conversely, the dangers are all too apparent: such as zero visibility in whiteouts, crevasses suddenly appearing in the ice pack, or electrical equipment failing in the extreme cold.

Equipment such as sextants and gyrocompasses used to be the essential navigational tools in these extreme environments. Using a sextant (in conjunction with an artificial horizon) is a good way to fix your position, but this requires good weather, an accurate chronometer and extensive knowledge on how to use navigational tables. These tools have been replaced today with the use of modern compasses and satnavs – both straightforward to use and the latter accurate in all weathers.

Specialist navigational equipment

Compass
Your compass must be balanced to compensate for **Magnetic Inclination** (see pp. 65–6) of the polar region you will be navigating and operate flawlessly in extreme cold. I recommend using the globally balanced Suunto M3 Global compass which I have used at –45° C – the notched bezel makes it easy to use while wearing gloves. I do not recommend mirrored compasses as these easily mist up and this condensation can then freeze. If you do not use this model, check the operational temperature range and inclination of your compass with the manufacturer.

Compass holder

This simple and lightweight compass holder allows you to travel and navigate without the need to constantly take it out of a pocket or pouch using your hands to hold the compass (see **www.snowsled.com**).

Satnav

Satnavs are not as valuable as elsewhere on the planet because they only work accurately at certain times of the day in these regions. This is because the orbital planes of all GNSS satellites have an angle of 55° with the equatorial plane, taking the satellite up and down between 55° north and 55° south. This means that near to the poles the satellites remain close to the horizon so coverage is limited to certain hours of the day – outside of these hours they either do not work or their accuracy should not be relied upon.

It is imperative that before going to these regions, you check the times of these windows of availability for your location and plan your satnav daily usage around. Bear in mind that recording of distance travelled, as well other trip data, cannot be relied upon.

At temperatures below –20° C, satnavs start to fail, usually with screen problems. For this reason I carry my satnav inside my jacket, sacrificing exact accuracy for practical usage and lower battery drain. When I wish to get a precise fix or set a bearing for an **Attack Point** I take it out of my jacket for a few minutes only. If I have set a new bearing, I transfer this to my compass for the duration of the leg. This short exposure of the unit to the cold prevents it succumbing to these severe temperatures. A must is that the unit can use lithium batteries (see **Battery Selection**) and operate in extreme temperatures.

The closer to the poles, and in extreme cold (≥–40° C) I navigate, I tend not use the more modern satnavs – since most do not provide accurate mapping for much of the region. The colour hi-resolution screens malfunction quickly in the cold and they do not have the facility to connect either an external antenna or external power-pack, allowing both the unit and extra batteries to be protected from the cold. The Garmin 60Csx has both of these features and has been my mainstay in these regions, when used in conjunction with my map and compass. I have used it at temperatures as low as –45° C.

Trip-wheel

In addition to both a spare compass and satnav if travelling by sled the simple, yet very effective, trip-wheel with an odometer can be used to estimate distance covered to augment the satnav's readings – GNSS does not show the true distance travelled over slopes. If necessary, a trip-wheel can be used in isolation for **Dead Reckoning**; this item should be considered as a failsafe.

Compass binoculars

I believe that these are an essential navigational tool in these environments. Many of these regions appear featureless, with vast level plains, so having the ability to identify distant features and take bearings on them very accurately is essential. I use the Steiner Commander 7x50XP to take bearings on major attack points then, while en route to it, I use my baseplate compass to stay on track. They are filled with nitrogen

and are therefore condensation-free during use (changing temperature and altitude causes this) and again I have used them at –45° C with no loss of function. In all-white environments, depth perception is reduced along with judgement of distance so the range-finding reticule is of enormous use.

Specialist techniques

Route planning
You cannot safely navigate under Arctic conditions without a route plan. This should be created prior to the expedition and yet must be detailed and flexible enough to adapt to changes in prevailing weather conditions and the fitness of the group. The environment you will be traversing can change dramatically throughout the year – for example, streams, rivers and deep soft snow may no longer be obstacles as they freeze. Glaciers mapped should be regarded with caution and it is important to know the speed at which the glacier is travelling (see **Mountain** section). Always make and take a copy of your route plan and store separately, ideally with another member of the group or, if travelling alone, use the ***Solo Rule***: 'one in rucksack; one on person'.

Bearings
There are many metal items that can severely affect the accuracy of your compass: snow mobiles, ski poles and ice axes for example. However, having to constantly dismount from your snow mobile, or place your ski poles behind you, is not an efficient use of time in this hostile environment – measures need to be taken to reduce the frequency of these stops. Fortunately, it is generally true in these regions that there is a constant clear view of the entire horizon, usually accompanied by clear skies and good visibility, providing the ideal conditions to use **Radial Arms** (pp. 146–7).

Keep walking poles away from compass when taking a bearing.

Care must also be taken in regions of localised magnetic variation, such as Ross Island in the Antarctic, where iron deposits in the local geology significantly influence compass readings. If your map states this variation, you can compensate for it on your compass. These differences can be very large and change significantly over short distances; for this reason it may be safer to use **Grid North** instead of **Magnetic North** or **True North**.

Be wary of taking bearings on features that are not immediately identifiable on your map – you could be sighting on an iceberg (the largest recorded was 168 m high, though 75 m would be a more usual height). Icebergs can become frozen into the icepack and so appear to be permanent land features – but will probably be drifting with the surrounding ice. The same is true of pressure ridges (up to 30 m) caused by the movement of the icepack.

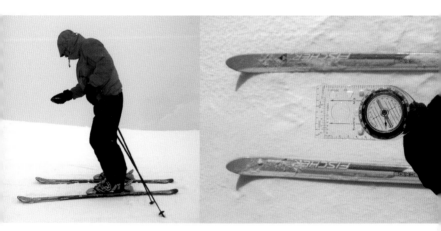

Ski alignment

When you have set the bearing on the compass you intend to follow, look through the compass' baseplate and align your skis with the direction of travel arrow. Whenever you stop, to quickly check that you are still on the intended bearing again, look through the baseplate and check your ski alignment.

Slope aspect

In the absence of features such as forests, foliage, rivers and streams, movement by terrain association becomes easier as observation is greatly enhanced.

Sometimes the combined effect of bright sunlight shining on snow can reduce contrast and makes reading the land angles over distance more difficult, so close, small changes in slope angle become important. Contrast can be enhanced with yellow tinted goggles. See **Slope Aspect** technique, pp. 154–7.

> ✖ **TIP:** The two other techniques regularly used are: **Dead Reckoning** (pp. 158–9) and **Back Markers** (p. 207).

Position marking

This technique should be used as a matter of course for back bearings. The variation on the standard method is to place a dark object on top of the snow mound you create (see **Position Marking**, p. 100)

Whiteouts

A whiteout is when you lose horizon and everything around you, including the sky, appears white, erasing all signs of shadow and definition. You lose perspective and cannot judge distance ... or, indeed, see anything at all. There are four types of whiteout.

Conditions and visibility can deteriorate rapidly: these two photos were taken two minutes apart!

- Arctic surroundings create unique whiteouts, usually in the springtime and often when the weather is calm and the visibility is excellent. When the sky is overcast, a thin layer of cloud diffuses the light, and can make the entire environment turn white when it is not snowing – these conditions can last for many hours.
- During a normal snowfall a sudden heavy flurry, albeit usually brief, can sometimes strike and create whiteout conditions.
- In blizzard conditions the volume of wind-blown snow, often mixed with ice particles, is so dense it is all you can see.
- On pitch-black nights even light falling snow can reflect the light from head torches and create a whiteout. Dimming your torch beam can reduce, yet not totally eliminate, this effect.

All whiteouts are dangerous – it is very easy to get disorientated, even within a couple of metres, and the risk of becoming confused and lost is very real. Stay put, if safe to do so, during whiteouts. The risks are high and in the absence of a horizon we easily lose our spatial orientation which can lead to vertigo.

If you **must** move, there are some measures you can employ to minimise this risk.

1 Before moving off, be absolutely certain that you have a fix and not an estimated position of your current location. If you are not, go through the **Relocation Procedure** and only then move off.

2 If you have one, use your satnav. It will tell you where you are and where to go, but it is important to stress that it does not tell you how to avoid snowdrifts, ice that has become open water, and avalanches.

3 Whiteouts are frequently a local phenomenon and related to a specific altitude so, if safe to do so, move to a lower position if it is not too far to travel. If your route was from a lower altitude, follow the old track as it may also give some kind of contrast in the snow.

4 If there is a forest nearby head for this; it will afford some protection from the weather too. If there is a **Handrail** nearby which is near impossible to miss, aim for it.

7 If you are using dogs, send one out ahead on a long leash/rope and be prepared to let it go if it falls over a steep edge: it is your life you are trying to preserve!

8 With the loss of an obvious event horizon, it can be difficult to assess how the ground lies ahead if you are moving up or down and depth perception is also difficult. Throwing a rope or a snowball out in front is a good method.

The snowball appears to:	The ground ahead is probably:
stick in mid-air	inclining upwards
land lower than your feet	inclining downwards
disappear	a direct drop – repeat in case you threw it too far and remember there may be a dangerous drop ahead of you

The most important thing to do is obtain a fix – not an EP – before the whiteout hits you. You will always have warning of it, even if it is only a few minutes.

Glacial navigation

All glaciers are continually moving masses of ice, water and debris – as a direct consequence maps of them rapidly become outdated, especially when crevasses (the main hazard encountered on glaciers) are marked. It is very important to remember that crevasses both appear and disappear. When moving over a glacier avoid the edges where crevasses are more common and frequent and there is the risk of avalanche from adjacent slopes. Use the **Boxing** technique (see pp. 143–5) when you encounter crevasses.

As a glacier retreats (as many are in our warming climate), the rock surfaces it was moving against are exposed and are usually polished smooth and very difficult to travel over safely. In addition, the process of retreat can mean that refuge huts, such as those found in Switzerland, that once stood beside a glacier can end up several metres above the glacier surface, making access difficult, if not impossible.

Environmental clues

Sastrugi
Sastrugi are sharp irregular grooves or ridges formed on the snow's surface by wind erosion and deposition and found in polar and temperate snow regions. They will run parallel to the prevailing wind – and will run on a bearing that you can use visually as you travel. This bearing needs to be taken standing.

1 Use a second compass which is marked to identify it as such. Point the direction of travel arrow towards an identifiable feature.

2 Ignore the compass needle.

3 Move your head immediately over the compass housing to avoid creating parallax.

4 With one hand hold the compass still so you can see through to the ground.

5 Rotate the bezel until the red orienting arrow runs parallel with the sastrugi. Check again that the compass is pointing exactly towards your identifiable feature and that the red arrow and the sastrugi are in perfect alignment.

Clouds
The surface of the earth's reflection on the bottom of clouds can indicate the type of terrain beneath them:

- **Black** – open water, snow-free ground and forestation
- **White** – snow fields and sea ice
- **Grey** – new ice
- **Mottled Grey** – pack ice or drifted snow.

Distant bearings

Frequently the nearest feature marked on your map will be some distance off and can only be clearly identified using the compass binoculars.

1 View the distant object through the binoculars.

2 Record its bearing.

3 Identify a feature nearer to you on the landscape in line with the feature you have identified – together they create both an attack point and a **Transit Line**.

4 As you navigate towards it and the distant feature becomes visible to the naked eye, you can use this transit line to check your bearing is correct.

Celestial navigation

Radial arms (snowmobile)
A simple yet highly functional adaptation of the **Radial Arms** (p. 144) can be employed.

1 Point the snowmobile in your intended direction of travel.

2 Mark the screen visor at the top in the centre with a line thick enough to create a shadow (an indelible alcohol marker pen works well).

3 Now mark the shadow cast either by the sun or moon onto the dash, this time with a pen line which can be removed. This is a quick visual reference which can easily be checked.

5 Repeat every 20 mins because in polar regions the sun traverses the sky more equally, as perceived on the ground, at 15°/hour.

↘ WARNING

Some snowmobiles come already equipped with a compass which has been corrected by the manufacturer for that machine and have inbuilt satnav or odometers. As a safety precaution you must still carry your own compass.

Night-time navigation
Clear open skies both during the day and at night make **Celestial Navigation** an essential tool when moving in this terrain (see pp. 77–85).

Moonlight and starlight on a clear night reflect off the snow, thus enabling you to employ daytime terrain association techniques with little difficulty. Even cloudy winter nights with snow cover can be brighter than clear moonlit summer nights in warmer climes, when the ground is dark and often covered with foliage.

Celestial navigation, employing both the use of the sun and star constellations, is particularly effective. Ursa Major (The Great Bear) is constantly above the horizon in the Arctic region and should be frequently referred to in confirming north. Conventional **Radial Arms** using both the sun and the moon are also highly successful.

→ At extreme temperatures (lower than -40C°) small bubbles may appear in your compass housing but these will not affect its performance. However, on the polar plateau at altitude, these bubbles can enlarge considerably to the point of making the compass inoperable.

→ If very near the magnetic poles, a special regional compass will be required.

→ Both Galileo and GLONASS intend to provide more satellite signals that will provide GNSS in extreme northern and southern latitudes.

→ Walking can take up to five times as long as it might under warmer conditions.

→ When giving a weather observation, wind direction must always be given in relation to grid north.

Maps

If travelling within 2,000 km of either pole the best maps to use are those which employ the **UPS Grid Reference System**. This is an accurate datum and grid references can be plotted in the same way as with the UTM grid reference system, with a slight difference to north and east references (see **Global Mapping Systems**, p. 42).

Aerial photographs generally should not be relied upon unless they are weeks old rather than months, never years, as the landscape is continually changing due to the weather.

Keep your maps in a waterproof case and carry a complete set of spare maps.

Near the ice break the landscape will be continually changing.

BAD WEATHER NAVIGATION

Immediately before a sudden change in the weather bringing a sandstorm, blizzard, whiteout or descending/ascending thick fog/mist, there are warning signs – learn to recognise these.

Ten golden rules for navigation in bad weather

1 Decide if you really need to move or can safely shelter until it passes.

2 Use shorter legs: every step is critical and this allows you to continually review your progress and be able to choose small features.

3 Use **Back Markers**.

4 **Aiming Off** is often safer than trying to find a single feature from a distance.

5 Anticipate what you are going to pass and use **Collecting Features**.

6 Identify **Catching Features** to stop you in safety.

7 Make use of **Outriggers** to determine contour slopes.

8 Use **Steering** or a **Three-Person Line Check**.

9 Keep the group together at a steady pace.

10 Concentrate and lead with calmness.

Two views on the same route: four hours apart!

Put your heads together, discuss and get a fix!

First Response

- Stop immediately.
- Take a compass bearing on your **Attack Point** or, if this is no longer visible, any other feature you would be confident is on your map.
- Get everybody together.
- Ask people to think what they have just seen while walking and add this information to your own recollection – discuss it.
- Relate this and your bearing, to your attack point and features on the map to get a fix.

Back markers

If alone, use a stick as the back marker or if with another person, use them (they do not need to be a navigator).

1 Select the bearing you are going to walk on. Ask the second person (your back marker) to stand still and agree your signalling.

2 Walk this bearing to within a safe limit of visibility. Stop, turn around and take a bearing on your back marker. Move sideways to correct your position if necessary.

3 Put both arms up in the air to indicate to the back marker they must now walk to you. The back marker also puts both arms in the air to acknowledge and walks to you.

Repeat the technique as required.

Steering

This requires two navigators: one follows behind as a back marker and corrects **Drift**.

1 Select the bearing you are going to walk on and both set your compasses to this bearing.

2 The first person, who should be the most accomplished navigator, leads off on this bearing and walks to within a safe limit of visibility, stops and turns around.

3 The second person takes a bearing on the lead person. If the leg has been correctly followed the rear person puts an arm up vertically.

4 If the course needs correcting the rear person indicates which direction the lead navigator must move using their right or left arm outstretched.

5 When the lead navigator is in the correct position the rear person again raises their arms. The lead navigator confirms acknowledgement by raising both arms vertically, upon which the rear person walks to the first.

Repeat the technique as required.

Three-person line check

You need a minimum of three people for this technique, two of whom must be navigators.

1 Select the bearing you are going to walk on and set your compasses to this bearing.

2 The first person, who should be the most accomplished navigator, leads off on this bearing.

3 The second person, who need not be a navigator, follows in the footsteps of the leader, walking 10–15 m behind.

4 The third person, who is a navigator, follows using their compass to check the bearing, looking ahead at the two walkers to see if they are in line with each other.

5 The leader needs to stop at regular distances, at which point the rest of the group should stop immediately.

6 If the leg has been correctly followed, the rear person puts an arm up vertically. If the course needs correcting, the rear person indicates which direction the lead navigator must move, using their right or left arm outstretched.

7 When the lead navigator is in the correct position, the rear person again raises their arms. The lead navigator confirms acknowledgement by raising both arms up, upon which the group start walking again.

Repeat the technique as required.

Pacing and timing

While these techniques are described separately in this manual, when used in combination they are a very powerful aid to navigation. When visibility is significantly reduced our pacing naturally changes, we tend to take shorter steps and increase our level of drift. An excellent technique to practise and help overcome these tendencies is to take bearings and use timing and pacing for a blindfolded circuit on a large, level playing field.

↘ EXPERT TIPS

→ Buy a book on how to interpret the weather.

→ Small objects can look much larger in snow, fog and mist, so be careful.

→ Look carefully for rocky features and triangulation points, cairns which stand out in the snow.

→ Throw snowballs ahead of you if the wind permits.

→ Snow lies at its thickest on the lee slope and in the depressions of hills so choose windward slopes and ridges to avoid deep snow if the wind is not strong.

→ In snow avoid slopes which are steeper than 30° as they are prone to avalanche.

→ Strong winds are tiring, blow you off course, drive snow, rain and hail into your face and reduce the weatherproofness of your clothes, so avoid walking straight into a headwind.

→ Keep off high ridges in strong winds. Wind accelerates up slopes and decelerates down slopes. Follow depressions to keep clear of the weather.

DESERT NAVIGATION

A third of the earth's land surface is occupied by desert. These regions have very little rainfall (< 250 mm/year).

There are four types of desert:

- Ice
- Sand dune
- Rocky plateau
- Mountain desert.

The world's largest desert is Antarctica and covered in **Navigation in Extreme Cold Environments.** This section deals with the more typically recognised deserts, which are dry, hot arid regions where vegetation is scattered, terrain is generally sand or rock, and the sky generally cloudless.

There are two general desert environments: broad basins between mountain ranges and large, flat open expanses where there are little or no obvious visible features or landmarks, often composed of repeating landforms such as sand dunes.

In the broad basins there are typically many terrain features, from mountain peaks to long ridges which you can use both as **Attack Points** and to take bearings from. Yet using these prominent features over great distances can easily introduce inaccuracy into your estimations, particularly when calculating a **Resection** (see pp. 125–6),

because one or two degrees' miscalculation is greatly amplified over such a distance. Therefore, frequent resections should be conducted and, where possible, accompanied by referencing close-in terrain features.

In large open expanses with few visual cues or where there is restricted visibility, such as a sandstorm, **Dead Reckoning** is the primary navigational tool.

Moving in the desert can often involve vast distances. Nevertheless, these distances are still broken down into relatively small navigational legs – micronavigation. However, legs in this terrain can be longer, up to 1,000 m, and for this reason bearings taken with your compass need to be much more accurate (see p. 12). This need for accuracy, combined with the fact that accurate sighting of distant features can be difficult, means that additional navigational equipment should be used.

Desert maps are often unreliable, and where the terrain is flat and open their contour intervals can be very large, so many of the relief features in-between are not detailed. Use more than one map of the region and augment them with up-to-date aerial photography, whether taken from a satellite and obtained from a resource such as Google Earth or Microsoft Virtual Earth, or from aircraft and purchased from the local government authority.

Specialist navigational equipment

Mirror compass

The Suunto MC-2 Global mirror compass is a sighting compass which is more accurate than a conventional baseplate compass, as it uses a sighting notch, yet it is just as reliable and rugged. It has a globally balanced needle and a declination adjustment system which is useful when working in areas of large magnetic variation.

Because features in the desert are often a substantial distance away, it is better to stand and take bearings rather than adopt the **Brace Position**, as you can see 30% further to the horizon standing (1.7 m) compared to the brace position (1.0 m).

Navigation logbook

A small notebook and pencil are all that is required to record headings, time, distance, bearings of landmarks and any significant features such as a branch or intersection of a track, habitation, or a water hole; it is an invaluable record of your journey for three reasons:

- Relocation – if you become lost or are unsure of your exact location this data will allow you to relocate. From the movement of shifting sands, large sand dunes can have moved or changed shape significantly and along with vehicles not needing to follow conventional routes, prominent tracks and trails can suddenly end or change heading unpredictably. Keeping notes of previously encountered branches in the track and landmarks, combined with how far back down the track they were, can indicate where you may have taken the wrong turning or made a wrong assumption about your route.
- Basis of **EP** – record from which dead reckoning position is deduced.
- Rescue – keeping accurate details of where villages, settlements, water holes and major tracks were last seen is critical if faced with an accident. Using the log you can accurately calculate practical walking distances, the potential visibility of rescue flares and make the decision on whether to continue, stop or go back.

Compass binoculars

Not only are distant features often difficult to see, shapes can be very deceptive in the desert and a pair of Steiner Commander binoculars, with a built-in compass, are an important additional navigational tool. Plus, they can also be employed during **Night Navigation**.

Satnav

With the large open skies of the desert both the number of available GNSS satellites and their signal strength is high and a handheld satnav is the most valuable navigation tool available. Check the IP rating of your intended handheld (see **Buying a GNSS Receiver/Satnav**) to ensure that it has full protection against the ingress of dust. I recommend IPX7 or better.

There are special considerations when using this technology in deserts.

1 Prior to commencing any trip, your planned route, potential escape routes, danger areas, nearest habitation and known water holes should be pre-loaded into all handsets.

2 Desert trips can often be for extended periods of time and small solar battery-charging units should be used with rechargeable batteries.

3 Handheld satnavs should not be exposed to direct sunlight in extreme temperatures for long periods of time, so the use of an external antenna should be considered.

4 Familiarise yourself with the use of the latitude and longitude coordinate system.

Specialist techniques

Route planning

Locate and mark refuges and water holes on your map and store them as waypoints in your satnav; a longer route that passes near to these can be much safer than the direct route across the middle of the desert. Familiarise yourself with the climatic changes in the region and local land phenomena: for example, some deserts have arroyos, also called a wash, which temporarily fill with water after either heavy rain or seasonally and can be tens of kilometres across. Make sure that you share the route with every member of the expedition, even if using a local guide, as in an accident, you must be able to describe exactly where you are and also how to escape.

Dead reckoning

Probably the single most important technique employed in desert navigation. Draw up a log of each leg length, bearing and time taken.

Transit lines

After taking a bearing on a distant feature, line it up with an intermediate feature both to use as an **Attack Point** and as a **Transit Line** to ensure a straight bearing is followed.

Markers

- Use your tracks created in the sand.
- When you either arrive at your attack point or stop part way along a leg.
- Turn 180° and check the bearing of your tracks.
- The white needle of your compass should point north on your bearing.
- If you have drift correct this and proceed onto the attack point.
- Alternatively, create a simple mound of sand onto which you must place a contrasting object, such as a stone, from which you can take a **Back Bearing** and therefore minimise the risk of creating parallel bearings.

Terrain association – dunes

The most prominent features in sandy deserts are the sand dunes. There are five principal types and these can be used to assist navigation. Knowing these and the direction of the prevailing wind you can utilise some of these dunes as **Radial Arms**, thereby always giving you a reference to the four cardinals of the compass.

Crescentic (barchan) dunes

The most common type of sand dune. As its name suggests, this dune is shaped like a crescent moon with points at each end. They are wider than they are long, being formed where there is a dominant prevailing wind – with a gentle slope upwind and a steep down slope downwind (sometimes referred to as the lee side of the wind). They run perpendicular to the direction of the prevailing wind. They move slowly (up to 20 m a year) and are usually 20–200 m long (the longest recorded was 3 km).

U-shaped (parabolic) dunes

Very common in coastal desert areas. Unlike crescentic dunes their crests point upwind, so the steep downslope faces the prevailing wind – the elongated arms of U-shaped dunes point in the direction that the prevailing wind is blowing from.

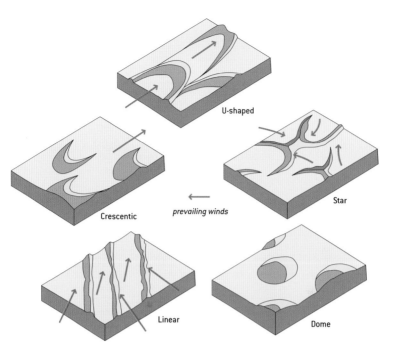

U-shaped

Star

prevailing winds

Crescentic

Linear

Dome

Star dunes

Pyramid-shaped with slip faces on three or more arms, occurring in areas of multidirectional winds – they cannot be used for direction-finding with a prevailing wind. However, they are very distinctive and often very tall (up to 500 m), so can be used as attack points, or (if marked on a map) for taking a bearing. They dominate areas of the Sahara, the Badain Jaran in China and are prevalent near mountain ranges.

Linear dunes

Long, snake-like ridges (up to 160 km) and generally form sets of parallel ridges separated by kilometres of gravel, rocks or sand. They form in bidirectional wind areas and the uniformity of their direction must not be relied upon.

Dome dunes

Shaped as their name suggests, they are usually of low height and are rare. Dome dunes are of little navigational use other than they can suggest the presence of a sand sea, as they occur at the far upwind margins of such terrain.

These five types of dune occur in three formations:

- simple dunes where only one type of dune is present
- compound dunes are large dunes on which smaller dunes of similar type sit
- complex dunes are combinations of two or more types of dune.

Simple dune formations are the most reliable indicators of direction as they represent a wind regime that has not changed in intensity or direction since the formation of the dune. Care has to be taken with compound and complex dunes formations as they often suggest that the intensity and direction of the wind has changed.

SPECIAL ENVIRONMENTS

➔ Contours and features, such as water holes, often change in the desert so only use the most recent topographic maps or aerial photographs.

➔ Most kinds of dunes are longer on the windward side, where the sand is pushed up the dune, with a shorter slip face in the lee of the wind.

➔ The valley or trough between dunes is called a slack.

➔ A dune field is an area covered by extensive sand dunes. Large dune fields are known as ergs.

Environmental clues

In addition to the signs detailed in the pages dedicated to **Environmental Navigation**, some techniques are unique to desert environments.

• The sun's heat distorts the air, which often shimmers above areas of hot sand. This does not tend to occur over areas of vegetation, so this phenomenon can be used to predict the presence such areas of land.
• Desert animals and birds move with regular patterns to the same water holes in the early morning and evening; make a note of where these are and try to relate them to your map.
• Stratum of rocks appear in all deserts, from small outcrops to massive ridges, and having a geological map of the region will give an Estimated Fix.

Night-time navigation

Soaring daytime temperatures in deserts can make movement across the land difficult and hazardous. Night navigation, where practical, in combination with **Celestial Navigation** makes movement more comfortable and navigation straightforward.

Radial arms

Sand dunes can undulate for hundreds of kilometres making identifying attack points and consequently maintaining a course difficult and sometimes impossible. As the desert sky is generally cloudless, it is an excellent environment to make use of both the sun and moon as a radial arm as you move from one viewpoint to another.

Celestial navigation

The desert is an ideal environment to use all of the celestial navigation techniques – with constant sunshine and clear skies with no light pollution at night.

Stereoscopic ranging

The absence of features on the land prevents comparison between the horizon and the skyline; this, combined with the mirages created by the hot dry air, means that distances can be substantially underestimated. Therefore frequent **Stereoscopic Ranging** should be employed. Where this is not possible, a general rule of thumb for estimating distance and time travel is to multiply your most conservative estimate by a factor of three.

Sandstorms

When the wind is strong enough sand is lifted from the desert surface and carried in the air. These storms can completely remove sand dunes and create new ones or move existing ones considerable distances. At their worst they can destroy encampments. **_Stay put when a sandstorm hits you_**. There are two principal types of sandstorm:

- Haboobs: common in the Sahara, which usually approach with little or no warning and take the form of a wall of sand, which can be as wide as 100 km. With winds up to 50 kph, movement within them is impossible. But they usually only last two to three hours, so taking cover and waiting for the storm to pass is the only option. At the tail of the storm, with appropriate respiratory and eye protection, movement is possible, but it is still easy to become disorientated and lost.
- Shamals: result from the NW winds that cross Iraq, Saudi Arabia and Turkey. They normally last three to five days, but can last for longer – the longest recorded lasted a week. The good news is that they are seasonal, so local knowledge will help you avoid them. Movement in this type of sandstorm when they are at their strongest (from spring to summer) is almost impossible. Outside this period, it is possible to move, again with the appropriate respiratory and eye protection and with the same caveat as above.

The most important thing to remember is that after the storm has passed, the landscape around you will have most likely changed, sometimes beyond recognition. Therefore, it is essential that as the storm approaches or just hits, you obtain a fix – not an EP.

FOREST NAVIGATION

A third of the earth's land surface is occupied by forests.

Forests range from areas of natural tree propagation, through man-made plantations (with regimented lines of similar or the same tree types) to dense, often impenetrable rainforest found mainly in the latitudes 10° north and south of the equator. Forests are differentiated from woodlands by the extent of canopy coverage. Woodland has a more continuously open canopy and trees spaced further apart, allowing more sunlight to penetrate. This section deals with both forest and woodland: rainforest and jungle navigation is covered in the next section.

All of these environments pose the same challenges to a greater or lesser degree to the navigator:

- restricted distance view, sometimes down to only a few metres
- obscured landscape panorama
- challenging terrain over which to maintain a constant pace
- difficulty in following a straight line of travel for any distance
- lower levels of light
- reduced GNSS signal strength
- incomplete or incorrect mapping – trees may have been felled or new trees planted
- incomplete or no mobile phone coverage.

As a direct result, it is easy to become disorientated in forests and potentially lost.

Specialist navigational equipment

- Satnav: In open forests where there are tracks and clearings satnav accuracy may be reduced but not significantly. In very dense forests where either the trees are very close together or the canopy thick, or both (especially in pine tree forests), an amplified antenna should be a considered, as inaccuracy both in height and location can be considerable due to lack of signal penetration and multipathing.
- Altimeter: if you are not using a satnav.
- Grease pencil/chinagraph.

Specialist techniques

Route planning
Do you need to go through the forest or can you use the edge of the forest as a **Handrail** and circumnavigate to your destination?

- What time of the year is it and what is the season? This will affect the vegetation: trees in full bloom with a thick canopy of leaves can reduce light levels. Heavy undergrowth, such as ferns and brackens, can make movement very demanding.

Large areas of water are easily identifiable on the map, but can be hazardous to travel next to – it is best to keep away from riverbanks, especially when the water is flowing fast.

- What time is sunrise and sunset? Make an allowance for shorter available hours of daylight as light levels will be lower in the forest at the beginning and end of the day.
- What are the weather conditions for the time of year? Small forest streams can become raging torrents of water and flat areas of land flood plains in the wet season.
- Aerial/satellite photography (say, from Google Earth) is very good for showing clearings and large tracks. But check how recent they are!
- Are there specialist maps available from forestry organisations? These may even detail the exact location of species type.
- If your planned journey is long are there any areas of refuge?
- Some tree plantations, particularly conifers where branches can come almost down to ground level, can be impenetrable.

Tracks

Most forests contain tracks, ranging from clearly defined man-made routes through the forest, to paths frequented by animals. The advantage of following tracks is that they are usually easier to move over and your range of vision is greater – this allows you to see the shape of the land ahead and carry out **Terrain Association**.

Position marking

Always mark on your map where you entered the forest on your laminated map and thereafter mark points where you have changed bearing or obtained a **Fix**. Join these marks up as you move, so you create a visual track record of where you have been.

Position marking

→ Older Ordnance Survey 1:25 000 maps used to have tracks marked, the new ones don't!

- Many, *but not all*, man-made tracks are marked on maps.
- Animal tracks often lead to and from water.

The hazard with following tracks is that they can differ from those marked on the map. New tracks are created and run into old tracks, so constant affirmation of direction of travel and position is essential. In low visibility, **Motorway Exit Syndrome** can easily occur. If you are following a track in poor visibility, avoid the edge of the track – if another track bears off, you may well follow this in error. Instead, use **Pacing** and **Timing**, walking in the *centre of the track*, to determine where you are and when you should reach a junction – at which point you either change direction or reaffirm your position.

Dead reckoning

It is difficult to maintain a constant pace, even in a straight line, in forests and for this reason make your navigational legs shorter than normal.

Radial arms

If the sun or moon are not clearly visible, observe your shadow and those of the trees around you and align your arm with these.

Altimeter

Where a forest is growing on an area where the land height changes, using an altimeter will greatly assist in determining a **Fix.** Confirm your location by searching on your map for the contour height at your location and make sure this tallies with your altimeter reading.

The wheel

Ideally performed when in a group – yet can still be an important aid when travelling solo (see **Jungle Navigation** below).

SPECIAL ENVIRONMENTS

JUNGLE NAVIGATION

Jungles are areas of tropical rainforests where the ground has been colonised by a dense, tangled growth of shrubs, trees and vines.

In jungles the annual minimum rainfall is 2.5 m, but is more usually greater than 5 m with high relative humidity, usually greater than 77%. They are all located in the intertropical convergence zone where the winds from the northern and southern hemispheres come together. They experience a monsoon season and a somewhat shorter drier season.

The areas of jungle in America, Africa and Asia all resemble one another in that many of the trees have trunks with no branches below 30 m. The topography varies considerably, often within the same jungle: in the Amazonian rainforest terrain ranges from the slopes, cliffs, hills and waterfalls of the Andean region to the heavily forested, relatively flat region that forms the majority of the Amazonian basin.

Where the leaf canopy is thin, the ground vegetation will be at its most dense – so, while available light to navigate will be high, the heavy layers of tangled vegetation can be impenetrable. This situation is often found at the peripheries where the primary jungle has fallen down or been cleared.

Conversely, where the leaf canopy is dense, available light at ground level (even at midday) can necessitate the use of a head torch – but the ground generally has a mossy cover and the only obstacles are fallen trees.

The most difficult navigational aspect is that it is almost impossible to move in a straight line, so specialist techniques have to be adopted to compensate for this.

> → A small detail which always makes people smile is that I carry a small pencil sharpener; I just find it a lot more straightforward to use than my knife!

Specialist navigational equipment

- 50 m of paracord
- waterproof grid paper and a grease pencil/chinagraph
- Suunto M-3 Globally balanced compass

- Satnav is of limited value but should be taken for when you are in clearings or at an elevated location with a clear view of the sky. Even with an amplified antenna, in the rainforests of Peru my unit averaged an accuracy of only 60 m, putting me inside an area of 150 m – which was unacceptable.

Specialist techniques

Route planning

These areas are not typically accurately mapped, if at all, as the leaf canopy makes aerial survey very difficult, so as many maps should be obtained and compared to each other as is practical. Route selection should be based on the contours and drainage pattern. In steep terrain, stick to the ridges to maximise movement and seek out clearings wherever possible.

Movement across the land will be slow, just how slow depends upon the density of the vegetation and the limits of observation, which can be only a few metres. Travelling between 200 m and 300 m per hour is realistic, not just because of the aforementioned obstacles but also due to exhaustion. The maximum is 1 km per hour.

An important consideration is the time of year you are intending to travel as during monsoon season regions can become hazardous, with large streams, rivers and deltas in flood. When in coastal areas, check inlet water and flood times and check contours to make sure you stay above these.

SPECIAL ENVIRONMENTS

Areas shown as cleared on maps can quickly become overgrown and impassable again and some swamps remain impassable all year round, particularly mangrove swamps, where the exposed roots of the trees are very difficult to traverse – here movement can be restricted to 100 m per hour.

The Wheel combined and **Dead Reckoning** are the two most reliable techniques for safe navigation in the jungle.

The wheel

This is primarily employed either to confirm your position, converting an EP into a fix, or if you are lost, to help establish your current location. Effectively, you are creating a mini map of an area covering 470 m^2

1 Secure the paracord to your rucksack and also to yourself. Using your compass cardinal north move out in as straight a line as possible. Count the metres on the paracord as you travel.

2 Use your grease pencil and grid paper to record any change in slope angle or features, such as streams or waterfalls against the number of metres away from your start.

3 At 50 m return to your starting point using the paracord to help direct you.

4 Now repeat this using the east, south and west cardinals of your compass, until you have collected enough information to help you arrive at a decision. If you are still unsure, use the NE, SE, SW, NW cardinals to augment your map.

5 In primary jungle and deep vegetation, repeat this procedure every 500 m or hourly, whichever is the shorter.

Dead reckoning

Somewhat difficult, though certainly not impossible. Very short legs are the key to success. Use if there is no obvious feature and visibility is limited to 10 m or less

1. Attach the paracord to another member of the party and also to yourself.

2. Send them forward up to the limit of visibility.

3. Use them as a marker You may find calling out easier than arm signals so remember your right is their left and *vice versa*.

4. Navigate to them on your predetermined bearing and continue past using them now as a back bearing.

It might not appear very quick, but in the jungle this technique is actually an efficient method of accurate movement.

Back snaps

Every time you make a significant change in bearing or come across a prominent feature, such as a river crossing or clearing, photograph it and note the direction you are facing and the grid reference: this is a record of what to see if you need to return along this route. Digital cameras are ideal for this.

Aiming off

If crossing a swamp is unavoidable, your route should be a straight line from a well-defined starting point and ending at a clearly identifiable **Attack Point**. Aim off to a prominent feature from which, when reached, you can turn left or right to locate your final attack point.

Terrain association

Slope angle, in addition to features such as streams, rivers and waterfalls, is central to jungle navigation and notes need to be made of every change and find, either using these to confirm your location on the map, or marking them on the map, if they are not already marked.

Navigation logbook

Continually record on paper the distance and bearing you have followed. If necessary you can plot these on your grid paper to show where you are in relation to your start and then transfer this information to your map.

Back snaps

Terrain association

Satnav

This is the most challenging environment for satnavs due to poor satellite signal reception. The dense tree canopy can shield much of the sky and consequently an amplified antenna is essential. Whenever you are in a clearing or at an elevated location with a clear view of the sky confirm your location and mark this postion on your map.

Jungle expeditions are frequently long in duration so take a good supply of lithium disposable batteries as they are both light and provide the most power.

Satnavs capable of displaying custom maps (see pp. 305–6) allow you incorporate recent aerial photography that can be used as maps with your location, track and intended route displayed on the photograph. Instead of downloading an exceptionally large file which the satnav's processor may find difficult to manage, it is better to download the area as a tile-set, where the relevant area is divided into multiple, separate images.

MOUNTAIN NAVIGATION

Scottish Mountain Rescue statistics show that the most common cause of mountain accidents, in a quarter of all call-outs, is poor navigation!

There is no universally accepted definition of a mountain – for this manual, we will take the definition to mean a topographic feature with a height over its base (not sea level) of greater than 1,500 m – they are steeper than hills and larger in landform. Two-thirds of Asia is covered by mountains, a fifth of Europe and America alike – but only 3% of Africa has mountain cover. Most of the world's rivers are fed from mountain sources.

We tend to think of mountains as static and not subject to change, yet there is a continuous process of change, altering the shape and characteristics of mountains due to a combination of glaciations, soil erosion, mechanical and chemical weathering (acid-rain) and global warming. So the mountain navigator must be aware that the map may have changed; everything from areas of forestation now impassable due to a previous avalanche, to cliff edges that have receded due to attrition.

Understanding the type of mountain range you will be visiting will help you predict the type of terrain, environment and prevailing weather conditions. There are five major types of mountains.

- Fold mountains are the most common type of mountains and examples include the European Alps and the Himalayas.
- Fault-block mountains, such as the Sierra Nevada mountains in North America, are formed when blocks of rock materials slide along faults in the earth's crust and the two types are lifted, which have very steep sides, and tilted, which has one steep side and gentle sloping side.
- Volcanic mountains are either active or dormant. They are usually of a uniform shape and slope – Mount Fuji, Japan, is a good example.

- Dome mountains are formed from molten rock when it spills out onto the earth's surface and spreads out, creating dome-shaped mountains such as the Navajo Mountain in Utah.
- Plateau mountains are formed by erosion and they usually occur near the fold mountain ranges. The Catskill Mountains, New York, are an example.

Some mountain ranges, such as the Rockies, are formed as a result of a combination of the above processes (i.e. folding, faulting and doming).

Specialist navigational equipment

Satnav

A satnav can be a life-saver in the mountains – I know this from personal experience!

I had ascended an exceptionally difficult ridge on a mountain called the Jungfrau, near to the Eiger in Switzerland, intending to meet the unique mountain train which travels up to the Jungfraujoch, the research station and observatory, that sits in the saddle between the Mönch and the Jungfrau for my descent. But, as often happens in the Bernese Alps, the weather suddenly changed, in this instance blowing in a snow

storm producing a whiteout. The conditions were treacherous and staying put was not an option. I was on a very narrow section of the ridge, where the wind threatened to blow me off the 3,520 m exposure and by this time I could only move on all fours with zero visibility. I had passed an area of safety 180 m beneath me and so decided to retrace my steps. I had with me an Etrex receiver and I set it to track back. This was my eureka moment with GNSS.

Altimeter

If you are not taking a satnav then a separate altimeter is essential.

Paracord

50 m of paracord: if visibility becomes an issue you can employ **The Wheel** technique.

Specialist techniques

Route planning

To determine what equipment you will require and how much you can achieve you must consider the following before your trip.

- Analyse carefully the actual distance you will travel, as this will be different to the measured map distance. Also assess the level of difficulty in climbing/descending.
- Plan and mark escape routes on your map, refuges such as mountain huts and make a note of what number to dial if you need MR. You should also input these as waypoints and routes if you are taking a satnav with you.
- What level of difficulty is the terrain you will be crossing? Steep rock can be difficult to navigate on and may require ropes or possibly climbing gear.
- How high are you travelling? Above 3,500 m oxygen levels are low enough to cause Acute Mountain Sickness (AMS). It affects over half of lowlanders who spend more than a few hours above this height and ranges from mild to severe. If untreated it can lead to High Altitude Cerebral Oedema (HACE) or High Altitude Pulmonary Oedema (HAPE) and both can be killers. In addition, the thin air provides less protection against solar radiation – sunblock is essential.
- How many hours of available light? Check sunrise and sunset times.
- What is the weather forecast and what is usual for the time of year? Do not rely upon a general synopsis, find an area-specific forecast (an example in the UK is the Mountain Weather Information Service at **www.mwis.org.uk**) and always be prepared for unusually favourable forecast weather to revert to seasonal type.

Terrain association and slope aspect

Frequently check the contours on your map and compare with the terrain in your immediate vicinity – continually use the confirmation part of the **Slope Aspect** technique as you travel through mountain ranges to confirm your location. The relocation part of this technique is also invaluable.

Determining elevation for height confirmation

Continually reaffirm location by comparing estimated position with the contour heights stated on the map. Also, whenever there is a critical turn or change in direction, especially where the path separates, mark its height using the altimeter in the satnav or with your altimeter. Then if you need to return via this route you have another reading to confirm your location.

You can also use your walking/ski poles to measure slope angle, as illustrated in the **Slope Aspect** technique.

Pacing and timing

The foreshortening effect (see p. 54) is the biggest consideration to factor in to your estimation of route distance in the mountains. Even with careful correction for slope angle, performing this technique on steep ground is difficult. However, over ground up to say a slope angle of 20° it can be an important technique to use both in both good and poor visibility.

GNSS

Often you will have a clear view of an open sky, especially near on at the summit, however on the way up and down you should be aware of multipathing – avoidance of this phenomenon is important (see pp. 285–6).

Environmental clues

In particular observe changes in temperature, vegetation and the wind direction. See **Environmental Navigation** section.

If you are navigating in a snow-covered mountain area, read the **Navigation in Extreme Cold Environment** section in conjunction with this one.

↘ EXPERT TIPS

→ The five most common mistakes made coming off a mountain.

1 Walking the wrong way off the summit by 180°: always check your bearing, preferably with another member of your group.

2 Miscalculating distance travelled: practise your timings and bearings in good weather.

3 High winds make it difficult to walk on a straight bearing: compensate for this.

4 Party separation spells disaster: always keep the group together.

5 Slips, trips and falls are more common on the descent than when going up. If the terrain is difficult, stop general chat and concentrate – and a slip can easily result in a fatal slide!

→ The five dangers of winter in the mountains.

1 Snow avalanche.

2 Snow cornices.

3 Shorter daylight hours.

4 Sudden weather change.

5 The cold.

SPECIAL ENVIRONMENTS

NIGHT-TIME NAVIGATION

The same techniques and practices that are used for daytime navigation are used in night-time navigation, with some subtle variations because judging distance, both visually and physically, changes in darkness.

There are many more hazards at night, where your visibility is limited, so practice during daylight is essential, as is preparing a comprehensive route plan including the lunar cycles.

Head torches have a limited beam both in terms of distance and spread, plus the artificial light often does not provide enough contrast to accurately read the land far ahead. **Pace Count (PC)** is lower as smaller steps are taken. With a limited view navigational legs (the distance between **Attack Points**) are shorter to reduce the margin for error. On starlit nights, with a full moon, movement can be conducted without using a head torch — it will only be required to read the map correctly.

Specialist navigational equipment

Head torch
Choose an LED head torch with a focusable beam for spotting features in the distance and which has different setting levels, so when used to read the map it can be set to low thus reducing the interruption of your natural night vision.

Satnav
An excellent tool for night-time navigation and some units have the facility not only to reduce the backlight but also display night colours to help you retain your night vision.

Night vision equipment
SAR professionals may wish to consider using this as it can greatly enhance night-time navigation. In line with all new technologies, it is becoming increasingly affordable.

Specialist techniques

Route planning
Firstly, if you are planning to be out all night ensure someone responsible knows where you are heading, where you are staying and when you expect to be back; ideally leave a paper copy of your route.

- Avoid steep ground, waterfalls, gullies and crags.
- Steer clear of boggy low-lying areas – they can be treacherous, especially when you cannot see the bog edges. Plan a route over dry, firm ground.
- Use **Linear Features** as much as possible, such as ridges in mountainous areas – so long as they are not too narrow so as to be dangerous. Other linear features such as paths, tracks and walls are ideal.
- Choose attack points that are both larger and nearer than you would pick during daylight hours.

→ Boost the luminous parts of your compass before departure under a bright light. During your journey expose it to your head torch for five minutes.

→ Always carry backup batteries and ideally a backup light.

• If you are planning a route which extends beyond the daylight hours, make sure you have plenty of opportunities to change the route part way through; escape routes are just alternative ways to shorten the journey.
• Know how to maximise and protect your vision.

Visual darkness adaptation
Our eyes adjust to the dark and can become 10,000 times more sensitive to light than at the start of the dark adaptation process. The brightness of the light and length of exposure will determine how long our eyes take to adapt to the dark.

Dark adaptation
After roughly 5–7 mins most people's natural night vision will be working at 30% of its capacity and after 30 mins full capacity – so, after turning off your torch, wait at least 7 mins before deciding if you are going to navigate using natural vision.

Protecting night vision
If bright light is unexpectedly introduced after you are dark-adapted, such as another person's head torch, quickly close one eye, preferably your dominant eye, to protect your dark adaptation. If necessary, cover it with your hand.

Increasing acuity
• Focus off centre: look 10° above, below, or to either side of an object rather than directly at it. This allows the peripheral vision to remain in contact with an object.
• Scanning the terrain: regard the object from right to left or from left to right using a very slow, scanning movement.

Head torch tenet
As a party you should all agree to reduce your head torch beam to minimum and point them down when facing each other or assembled as a group.

• To protect your eyes from your own head torch, either mount it onto your rucksack shoulder strap or wear a peaked cap underneath it.
• Consider using a blue or red filter over the lens; both help protect your natural night vision.

Pacing and timing
This is the mainstay of night-time navigation. A night-time **Pace Count** should be obtained using a predetermined 100 m course in the dark, in addition to your daytime PC. You will also tend to travel slower so take account of this when using pacing combined with timing.

Collecting features

The more features that you can tick off along your route the easier it will be to stay on the right course. Contours take on a whole new perspective in night-time navigation as there is less detail visible in the landscape due to the limits of your torch and lack of contrast so local contours describe the terrain in great detail. If you see car headlights look for a road/track in that area to confirm you location.

Features

Preferably, select prominent features which have a catching feature beyond them or a confirmation feature en route. Navigate to them using attack points which are less than 100 m apart and easy to see. Only ever navigate towards a light if you are certain what it is. Cars move off, light go out. Towns and built-up areas are much safer, as are lighthouse lanterns.

Handrails

It is often much more straightforward to stick to a track, beside a wall or stream, or any other linear feature; these may not be the most direct route but will be the safest and easiest to follow.

Aiming off

As visibility is reduced this may add slightly to the length of your journey yet will ensure you reach your attack point.

Terrain association

As you travel across the ground mentally note changes in the slope angle and aspect and relate these to your map.

Thumbing the map

Become much more aware of changes in your environment as you thumb your map noting the smallest of features and collecting them as you travel.

Markers

If travelling as a party you can lead out in front and have another member of the group use the same bearing to check that you stay on track and do not create a parallel error.

Celestial navigation

Refer in detail to the **Celestial Navigation** (pp. 69–85) section of this book.

Environmental navigation

At night, forests can give off a very strong scent so take into account the direction of the wind and what you can smell on the air.

SHORELINE NAVIGATION

Shoreline navigation refers to where the land meets oceans, seas, sea lochs, river estuaries and freshwater lakes and rivers.

The shoreline can be a very hazardous natural environment to traverse, where cliffs and ravines pose obvious threats, yet most critically land disappears when tides come in. For this reason an understanding of tides, tide tables and the ways high tide and low tide are depicted on maps is crucial.

Tides

All coastal regions have an area of land called the **Intertidal Zone**, commonly known as the foreshore, and this strip of seashore represents the greatest danger to safe navigation on the shoreline as it is submerged at high tide (HT) and exposed at low tide (LT). The principal risks in tidal areas are:

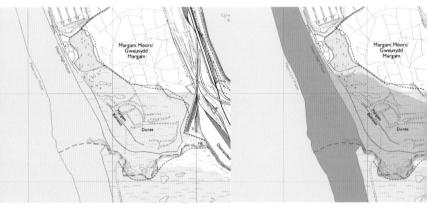

Bear in mind that during 'spring' tides, high tide will often flood a lot further inland (paler blue area) than the Mean High Water marked on your map.

- becoming cut off or trapped by a rising tide
- being dragged out to sea by the undertow even in very shallow water
- water levels rising very rapidly to quickly take you out of your depth
- inlets, large expanses of wide sands and coves becoming potential traps.

The difference in height between HT and LT is called the **Tidal Range**. This height difference varies from as little as a few centimetres in areas of the Baltic, Caspian Sea, Caribbean and Mediterranean up to 14 m in some areas of Britain such as the Bristol Channel. The largest tidal range in the world is found in Canada at the Bay of Fundy

The ever-shifting shoreline landscape of the Bay of Arcachon in France.

where it is 16 m. While this height difference is important its real significance is in the width of the intertidal zone it creates, which in some areas can be hundreds of metres, even kilometres wide.

Tidal changes can also reach far upstream in rivers. On the river Thames in England, tidal changes occur as far as Teddington Lock, 100 km upstream from the sea, and on the river Seine in France at Rouen, 100 km inland, the tide can rise and fall by as much as 4 m. The tidal limit on the Hudson River in the USA is the federal dam at Troy, 260 km upstream from New York!

Lastly, tides change sand landscapes over time so maps need to be as up to date as possible. The swirling sand banks that shape-shift around the Bay of Arcachon on the Atlantic coast of France are dramatically changing the shape of the intertidal zone.

Tide types
There are three types of tide:

- *Semi-Diurnal (same height)* mainly occur in the Mediterranean and around the British Isles; there are two full tides in just over a 24-hour period – a tide coming in taking 6 hrs 12 min to rise and a tide going out taking 6 hrs 12 min to fall; high tide every 12 hrs 25 min; the greatest flow of water and the fastest rise of the tide is in the middle two hours – half the rise or fall will occur in these two hours.
- *Semi-Diurnal (mixed height)* mainly occur in western USA and Canada and have exactly the same time intervals as above; the two full tides within the 24 hr period are of different heights.
- *Diurnal* mainly occur in the Tropics; one tide a day; tidal range is very small.

Tide tables
As the moon orbits the earth its gravitational pull makes the surface of the sea rise and fall. The sun also affects the tides but less because it is much further away but when it pulls in the same direction as the moon it increases the effect of the moon and this

causes the highest tides – spring tides. When the sun and moon are at right angles to each other they create the neap tides. These both occur twice a month. ***Spring tides have nothing to do with the season of spring!***

Because astronomers can very accurately predict the movements of the sun and moon the heights and the times of tides can also be predicted. Local historical observations of the area are added to these computations to produce tide tables, but they are only forecasts and can change.

The single biggest factor is the wind, which can dramatically affect sea levels with storm surges creating long-period waves which give higher and lower than predicted levels. Atmospheric pressure effects can significantly alter the times and/or heights of the observed tide. So, when using these tables, a margin of error should be built into calculations when planning your route.

As mentioned above, tides also vary considerably from place to place. So while the majority of the Mediterranean has little noticeable tide there is a 1 m tidal range both around Gibraltar and Venice – and the largest tide in the Mediterranean is at Sfax, Tunisia, in the Gulf of Gabes with a 1.5 m tidal range. So local tide tables are vital.

As with all navigation your greatest asset is your common sense so, in addition to carrying tide tables, visual monitoring of the tide is a must.

Ideally contact the nearest coastguard or harbour master's office at a port near to where you intend to navigate. There are also several websites (listed below). Often local fishing tackle shops will distribute tide tables and they can also be displayed on boards by the sea. Whichever source you use check that the day, year and location exactly match your intended time and place of travel because tides continually change.

Using tide tables

In England and Wales, mean high water (MHW) is the high-water mark of an average tide and mean low water (MLW) is the low-water mark of an average tide.

However, in Scotland, mean high water springs (MHWS) are used and are the high-water mark of an average spring tide and (MLWS) the low-water mark. Spring tides result in higher-than-average high tides and lower-than-average low tides – so at times of the tidal month other than springs, the tidal range marked on the map will be overstated.

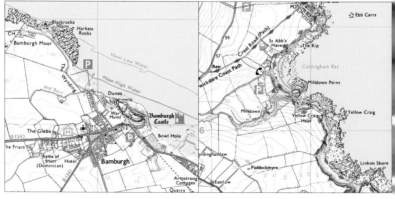

The shoreline at Bamburgh in England (above left) and Coldingham in Scotland (above right), showing the different representation of tidal range in the two countries. Also note that UK OS maps have a different contour interval at the coast – 5 m rather than 10 m.

The dynamic shoreline

Tides change the navigable land you can safely cross at any particular time, in addition features along the shoreline can change significantly. Coastal erosion in some parts of the world is both rapid and severe and substantial landslides can dramatically change the shape of the shoreline. Even man-made structures can move or be completely washed away so their location on a map may be misleading ... if they are marked at all! Take extra care when navigating to these or using them to determine your location.

Specialist navigational equipment

Compass binoculars

Use these to take bearings accurately and estimate distance. At sea level obvious features can be very difficult to find and for this reason a pair of good-quality binoculars with a built-in compass such as Steiner Commander 7x50XP will greatly improve your navigation. If not, ordinary 10 x 32 binoculars will make navigation easier and safer.

At sea level obvious features can be difficult to find – use compass binoculars to take bearings accurately and estimate distance.

Satnav

At sea a handheld satnav has perfect, large open skies for satellite acquisitions; however on the shoreline, and especially at sea level, the skyline can be obscured – sometimes as much as half the sky if you are next to cliffs. Also at the foot of cliffs **Multipathing** (see pp. 285–6) can affect the satnav. Therefore accuracy levels can be low and as a result your position should always be confirmed using a map.

There are special considerations when using this technology on the shoreline:

- prior to commencing any trip your planned route, potential escape routes, danger areas such as sinking sands should be pre-loaded into all handsets
- if your unit is rated less than IP65 (see **Buying a GNSS Reciever/Satnav**, p. 266) store it in a waterproof plastic bag and always secure it to your person using a lanyard.

Specialist techniques

Route planning
- Acquire the tide table before you leave.
- Follow the nearest contour above the highest tide stated in your table and make sure your route never reaches or goes below this level near the specified time. Repeat this for the low-tide mark.
- Draw on your map your safety margin for the tides and beside each note the tide times for that specific day of travel.
- Always plan potential escape routes should the tide come in.
- Ideally obtain the heights of large man-made objects on or possibly visible from your route such as lighthouses, power-plant cooling towers and chimneys, power transmission lines/pylons; these are all often situated by the coast. The internet is a good source for this information or perhaps call the service provider themselves.

Calculating tidal safety margin

1 Using your compass ruler, or roamer, measure the distance of the tidal range. Look carefully at the fall of the land (slope angle).

↘ EXPERT TIPS

→ On sandy beaches tidal currents running parallel to the beach and then out to sea often occur even when the tide is coming in (rising).

→ Flotsam and jetsam are usually found at the level of the last high tide.

→ Buoys and other light sources at sea should be generally ignored unless you are proficient in maritime navigation and have maritime navigational charts with you – it is easy to confuse one for another.

2 If the land rises steeply from the sea and the tidal range is low add 10% safety margin inland from both the low-tide mark and the high.

3 If the land is relatively level and the tidal range high allow a 50% safety margin.

Estimating distance using the range-finding reticule.

Lighthouses are excellent for estimating distance as they are often tall and highly visible. They can also be used as direct bearings or for a resection to determine your location. Alternatively if you are a sea level and can clearly see through the binoculars the top of a hill, possibly a triangulation point, which is clearly marked on your map you can estimate your distance from it by:

$$\text{Distance} = \frac{\text{Object known height}}{\text{Object size read}} \times 1,000 \text{ m}$$

See section on **Compass Binoculars** (p. 63) for some detail.

Stereoscopic ranging

Sterepscopic ranging on known coastal feature.

Cliff aspect

This specialist technique for position confirmation and relocation is covered fully in the **Techniques** section (pp. 174–5).

Transit lines

In shoreline navigation transit lines can be used to accurately fix your position.

Clear vistas, open water, land shape and mapped features help with the accuracy of shoreline transits.

Lighthouses

As their name implies they have the added advantage of being clearly visible at night. All lighthouses are identifiable by a unique combination of flash duration, frequency and interval between flashes. For example the harbour lighthouse at Penzance in Cornwall, England, flashes every 2 seconds, while Newlyn harbour light, 1.9 km away, flashes every 5 seconds. See **micronavigation.com** for access UK lighthouse information at Trinity House (England) and the Northern Lighthouse Board (Scotland).

Environmental clues

In addition to the signs detailed in the pages dedicated to **Environmental Navigation**, some techniques are particular to shoreline environments.

On sunny days the land heats quicker than the sea creating a breeze which blows off the sea and towards the land during the day; at night-time this process is reversed. Importantly in this weather system the clouds will probably be travelling in the opposite direction to the wind at ground level so ignore them.

Rock type can vary dramatically and is exposed where the sea meets the land. Mark on your map different geological features along your route; these are easily found on the internet. Learn to differentiate simple ones such as sandstone and granite and look at the stratums of rocks from small outcrops to massive cliffs. Identifying these rocks will give you an **EP**.

↘ TIDE TABLES

→ National Government Tide Tables (Home page given as individual pages subject to change). United States of America http://www.noaa.gov/ Canada http://www.shc.gc.ca UK http://easytide.ukho.gov.uk France http://www.shom.fr/ Germany www.bsh.de Australia, South Pacific and Antarctica http://www.bom.gov.au/index.shtml New Zealand www.hrdro.linz.gov.nz

URBAN NAVIGATION

When navigating in urban places, it is man-made features, such as roads, railways, bridges and buildings that become important, while terrain and vegetation become less useful.

For these reasons, while it is possible to navigate an urban environment with a 1:25 000 topographic map, it is much more straightforward if you use a smaller scale map such as Ordnance Survey Street Map (1:10 000). These maps are in the public domain as open data and as such can be downloaded and used by anyone. You can also obtain specialist municipal street maps which contain street names. In addition, they tend to be larger scale, which means more detail – and they are often more up to date. There are specialist map-makers and town and city planning offices can be a good source of urban maps – or they will be able to refer you on to contemporary urban map publishers.

Other sources of mapping include:

- The internet is now a good source of contemporary urban mapping (such as OS Street View in the UK and Google maps in most countries) and when used in conjunction with your satnav can be a powerful navigational tool.
- Aerial/satellite photography maps (which can be set to be overlaid with mapping data) can be downloaded and printed from Google Earth and Microsoft Virtual Earth – although, bear in mind that these images may be several years out of date.
- It may on occasion be necessary to create a sketch map as you navigate, for example in a newly developed residential area or after a major incident.

OS Street View and Google Earth with mapping overlay.

Specialist navigational equipment

Satnav

GNSS-enabled mobile phones come into their own in urban environments and can be used as the primary navigational tool. Many augment their location data and maps using Wi-Fi and cellular transmissions. Handheld stand-alone satnavs are equally good but in different ways, usually providing better batter life and features such as Trackback and a greater capacity to store routes and waypoints.

↘ EXPERT TIPS

→ Taking a bearing on high building – if you cannot see the base predict where it is and take your bearing in this direction to avoid tilting your compass which can lock the needle.

→ Successful navigation always requires concentration so also be very alert to the hazards of traffic.

Specialist techniques

Features

Building styles and sizes – specific areas, particularly in cities, will have similar architecture which was contemporary at the time of their construction and can often quite clearly define the area of the city.

- Urban geography of industrial buildings – invariably industry is built on the outskirts of town and cities but can over the years become a more intrinsic component of the town/city through urban crawl. Industry was often based near waterways so look for canals and viaducts.
- Residential housing – again will have been built in pockets with often the more affluent, ostentatious houses sited near parks or open land.
- Market districts – tend to be central and with roads that lead directly into them and out of the town/city.
- Man-made transportation features other than streets and roads – railways, canals and tramlines. Railways and canals will always lead out of the area and make good linear features which can act as catching features or transit lines.
- Terrain features located within the built-up area, particularly contours which can be related to the topographic map using aspect of slope, transits etc.
- Street signs – check how they are applied in this area; sometimes they can be placed on the corner of a building referring to the street around the corner.

Keeping on track

Because of the amount of navigational information your brain will have to assimilate the less you need to refer to your compass the better.

- At the beginning of the journey find a prominent building that you are confident you will be able to see throughout your journey, take a bearing on it and create a **Baseline**.
- Use the celestial navigation technique **Radial Arms** (pp. 146–7) to help maintain a bearing.
- Use television satellite dishes on buildings as a type of radial arm. They point to geostationary satellites that sit above the equator and are usually all aligned in the same direction. When first arriving at an urban area confirm which way they are pointing: align yourself directly in front of one, take a bearing from it.
- Use **Reverse Dead Reckoning** (p. 157).
- Be aware of unique features, such as a monument or a statue, read its plaque for a name if you have time and record this with your reverse dead reckoning notes.

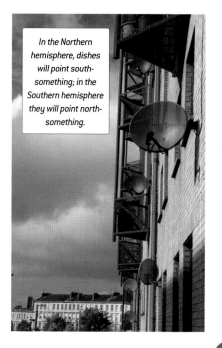

In the Northern hemisphere, dishes will point south-something; in the Southern hemisphere they will point north-something.

Add a catching feature, such as a road or railway line behind your baseline.

- Continually mark your position on the map.
- Whenever you have a vantage point with a view of two or more known features shown on your map, use a **Resection** (pp. 125–6) to get a **Fix**.
- Work hard to develop a mental map of the entire area. This will allow you to navigate over multiple routes to any location. It will also help protect you against getting lost whenever you miss a turn or are forced off the planned route by obstacles.
- Stop and ask people if you are unsure of your exact location and always either confirm this on your map or with another person.
- Make short summary notes during the route: e.g. *'5 min heading W on main road crossed railway bridge, 1st right on road following railway S, 10 min entered park area, RV @ cafe SE edge of park.'*
- If on a SAR mission offload as much of the thinking as possible to your Control so that you can focus on the execution of the current task.
- If following a predetermined route, identify **Collecting Features** and create a **Catching Feature** prior to setting off.

Satnav

The application of GNSS in urban environments has been the principal focus of manufacturers developing this technology for the consumer – primarily for use in cars and now pedestrians in conjunction with their mobile phones. This, combined with massively increased satellite performance, has yielded units which provide remarkable accuracy in urban locations and are an invaluable asset in urban navigation.

TrackBack

This is a good way of getting back to your car or hotel, albeit by the long way if you have wandered about! For SAR teams this provides the ability to accurately record exactly where you have been and if necessary, retrace this track. This information would then be given to the Search Manager who is coordinating the mission.

Waypoints

Waypoints can be marked anywhere for recording anything from a postbox to the exact location and time a casualty/item was found. Each mark is unique and forms contemporaneous evidence and should always be given to the Search Manager who can also assign them to photographs and notes in **Digital Mapping**.

Waypoints, routes and tracks can all be uploaded onto digital mapping.

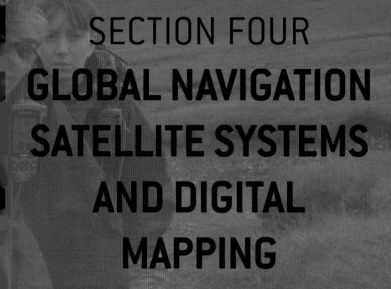

SECTION FOUR
GLOBAL NAVIGATION SATELLITE SYSTEMS AND DIGITAL MAPPING

UNDERSTANDING GNSS

At first glance this seems a pretty daunting section, designed for intrepid explorers and professional SAR teams.

Introduction

Global Navigation Satellite Systems (GNSS) are a part of our everyday lives, used in everything from domestic cars, to the navigation systems on all commercial and military aircraft and merchant ships. GNSS is also used to broadcast our location, with applications such as Google Latitude letting us see where friends and family are, and the European Commission's *eCall* system, where a car will automatically send its position following an accident to a call centre responsible for answering calls for police, fire departments and ambulance services. Many modern cameras have inbuilt GNSS and attach the shooting location to a photograph automatically.

'The evolution of the oldest of all navigational techniques, Celestial Navigation, has paved the way for a new revolution, where today we now launch our own celestial bodies to navigate by ... and it is called GNSS.'
Dave 'Heavy' Whalley, 2010

It is even used for 'just-in-time' manufacturing, mining, road building, farming — the soil in the field behind my house has been analysed and when the tractor is spreading fertiliser on it using GNSS positioning, different areas receive different amounts of fertiliser depending on the requirements mapped during the soil analysis!

Less well known are the hidden applications of GNSS, such as the high-precision timing it provides to keep our telephone networks, the internet, banking transactions and even our power grids online. GNSS has become part of the fabric of our society and without it many vital services could no longer function.

Basic units can be bought in supermarkets with more advanced equipment, such as heads-up display in ski goggles, available from specialist sports retailers.

All satnavs have the same basic functions, so gaining an understanding of these will allow you to get the best from the technology no matter what its application.

What is GNSS?

The Global Navigational Satellite Systems allow enabled receivers to determine location and height to within a few metres anywhere on the earth's surface, both on land and at sea (plus in the air), using microwave radio signals transmitted by satellites in orbit around the earth.

GNSS and GPS confusion?

Hoover is the name of a company, yet it has become synonymous with the vacuum cleaner. Global Positioning System (GPS) is the name of the American global navigational satellite system maintained by the US government, yet it has incorrectly become synonymous with similar systems realised by other countries. The European Union, China and Russia already have their own systems. The correct term given to all of these systems, including the American GPS, is GNSS and this is how all systems will be referred to throughout this manual.

The American GPS was originally created exclusively for military use and its signals were encrypted. This is still the partly true: the military are able to receive an encrypted signal called Precise Positioning Service (PPS) while civilians receive the Standard Positioning Service (SPS – also sometimes called coarse acquisition code or C/A code) which is not as accurate as the military PPS. It is generally accepted that the reason many other countries developed their own GNSS was because of the possibility that the US Administration could, if they so desired, encrypt all their signals at any point in the future. When all currently planned GNSS systems are deployed, the general public will have access to more than 100 satellites, which will significantly improve all the aspects of positioning and availability of the signals.

History
The Motorola company introduced the first mobile phone for general use in 1983, and the first SMS text message wasn't sent until 1997 – and yet by 2010 there were some 5 billion mobile phones in use around the globe. Within 25 years of its introduction this new technology has rapidly gained global acceptance – the mobile phone is now considered an everyday essential item. In addition rapid technological development and consumer demand has driven improved reliability, ease of user interface and affordability. Modern smartphones are essentially small personal computers that increasingly include GNSS technology.

Like the mobile phone revolution, the GNSS revolution is a great example of increasing features and decreasing costs, growing user-friendliness and all at a pace even faster than that of the mobile phone – 15 years:

Country	Name of GNSS	Name of SBAS	Planned full operation	No. of satellites	Civilian Accuracy	Military Accuracy
Global Navigation Satellite Systems (GNSS)						
USA	GPS	WAAS	1995	24	2.76 m	30 cm
Russia	GLONASS	SDCM	2011	24	2.8 m	5 m
EU	GALILEO	EGNOS	2013	30	2 m	10 cm*
China	COMPASS	SNAS	2020	35	3 m	70 cm
Regional satellite systems that will integrate with GNSS						
India	IRNSS	GAGAN	2012	7	20 m	–
France	DORIS	–	2012	4	15 m	10 m
Japan	QZSS	MSAS	2014	3	5 m	5 m

Prior to being fully operational all systems provide coverage to a greater or lesser extent. All SBAS systems work with all GNSS. All GNSS should work with each other (dependent upon all governments agreeing).
* Galileo system has a higher accuracy code available commercially

- **1972** The US GNSS system GPS (originally NAVSTAR) given the green light to start.
- **1995** The US GPS declared fully operational.
- **1997** Magellan introduces the first handheld GPS receiver priced under US$100.
- **1998** Garmin introduces the very popular Etrex handheld GPS for the outdoors.
- **2000** President Bill Clinton announced the removal of Selective Availability (SA) giving the public access to GPS.
- **2005** Google Earth and Microsoft Virtual Earth launched where GPS tracks can be superimposed onto aerial photographs of the world.
- **2006** All major car manufacturers offer GPS in their cars.
- **2007** The United States Department of Defence announced that future GPS satellites will not be capable of implementing SA and eventually making the policy permanent.
- **2008** The Civil Aviation Authority (CAA) allows non-precision GPS approaches for General Aviation aircraft in the UK.
- **2011** All commercial aircraft authorised to use Europe's EGNOS signal to land.

How GNSS works

As with all technologies it is easy to over-complicate the explanation of how they work. I have been fortunate to work with government agencies that both control and develop the satellites, which are at the core of any GNSS, along with numerous manufacturers of both civilian and military handheld units, testing and evaluating them throughout the world. The short outline given here is only what you need to know to use your handheld GNSS device – if you wish to learn more about this technology, including satellite types and the current status of the constellations, including orbit data and resources on active GNSS satellites, visit my website (**micronavigation.com**).

Your handheld satnav receives signals, just like a radio, from satellites which orbit the earth. The satellites transmit high-frequency low-powered radio signals which travel by line of sight and will pass through all weather systems and materials such as glass or plastic but will not transmit through most solid objects like buildings, mountains or water ... or indeed you!

From these signals your satnav can calculate where each satellite is, and then it simply triangulates your location from the known positions of these satellites, in much the same way you do when using a compass to triangulate (a **Resection**) and displays this as a **Grid Reference**, a mark on its electronic map, or both.

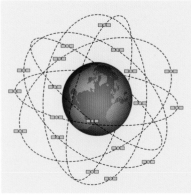 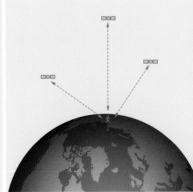

→ The more open sky your receiver can see the greater its acquisition of satellites and therefore the more accurate it will be.

→ As with a computer, update your receiver's software regularly: I do this monthly.

Once your position has been determined, the satnav can calculate other information, such as speed, bearing, track, trip distance, distance to destination, sunrise and sunset time and more.

Satellite Based Augmentation Systems (SBAS)

A series of ground stations are located at very accurately surveyed points on the surface of the earth. These take measurements of the GNSS satellites, check for errors and then transmit the corrected orbit data to geostationary satellites (satellites that orbit the earth in a fixed location), which in turn broadcast to the end users – that's you!

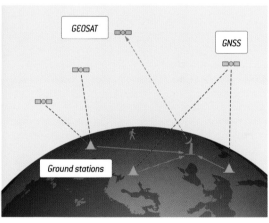

Ground stations receive signals from all visible satellites and calculate their errors plus the time delays due to the ionosphere. The ground stations then transmit corrected orbit data for these satellites via a central processing centre to the GEOSAT.

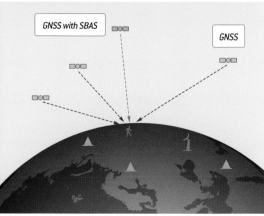

The user's satnav receives signals from the visible satellites plus the corrected data for these satellites and the ionosphere delay from the GEOSAT. It then uses all of this information to calculate your position.

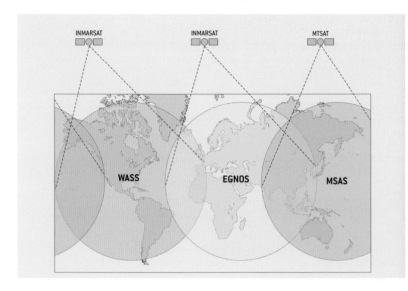

Each GNSS has its own SBAS: the European Galileo has EGNOS, Japan's QZSS has MSAS and the American GPS has WAAS. They are designed to be regional but can actucally be picked up across wide areas of the globe – parts of Europe benefit from the American WAAS and vice versa. All GNSS systems transmit information on the same frequency, so that the SBAS from the American GPS can be used by the European Galileo and so forth.

GNSS chip design

Satnavs process the information they receive from satellites to determine location, elevation, current speed and direction using their silicon chips – in practice they are miniaturised specialist computers. And as with computer chips, there have been incredible advances in GNSS chip technology and design.

High-sensitivity chipsets allow users to track more satellites in more challenging environments than ever imagined and two major developments to be aware of are *predictive ephemeris* and *dead reckoning*.

Predictive ephemeris

Every GNSS satellite continuously broadcasts a navigation message in three parts:

- the week number and very precise time within the week, as well as the data about the health of the satellite
- the ephemeris – this provides the precise orbit data for the satellite
- the almanac – this contains coarse orbit and status information for all satellites in the network as well as data related to error correction.

✖ **QUIRKY FACT:** US manufacturers cannot export a GPS receiver unless the receiver contains limits restricting it from functioning when it is simultaneously at an altitude above 18 km and travelling at over 515 m/s so that they cannot be used in ballistic missiles.

Predictive ephemeris enables the behaviour of GNSS satellites to be modelled based on broadcast ephemeris readings for accurate prediction of satellite positions. A single broadcast ephemeris reading allows the accurate prediction of satellite orbits, which are developed and refined every time new broadcast ephemeris is obtained and can work up to seven days ahead. So if you obtain a poor signal, no signal at all, or turn the unit off – when you turn it back on, or are back in an area with good reception, the device has a head start by knowing where it should be looking for the satellites. The result is that you can re-establish your location in as little as 3.5 seconds instead of up to 10 mins and predititve information can be relevant for up to seven days.

Dead reckoning

One of the most important requirements of your satnav is to maintain accurate positioning information – often in the most challenging environments. Chips have now been designed which have motion sensors to create a dead-reckoning solution. Put simply, when you are receiving only degraded satellite signals, or none at all, these chips measure how fast you are travelling and in which direction to estimate where you are at any moment based on your last measure coordinates. These measurements validate and correct GNSS fixes during signal blockage and multipath conditions. Chip manufacturers claim these systems will eventually make GNSS reliable inside buildings, in high-rise urban environments and challenging natural environments (such as mountain valleys and canyons).

These major innovations, combined with the increased numbers of satellites that can be tracked and processed by the chips and the lower power consumption of the receivers, will continue to improve both the accuracy and reliability of GNSS. Robust, reliable and accurate basic handheld units can now be purchased for little more than the cost of a compass and the technology will continue to revolutionise our lives.

Why use GNSS?

Handheld satnavs are the most powerful navigational tool available to the land navigator and have become indispensable for all serious outdoor navigators. There are no subscription fees or set-up charges – it is completely free to use!

When I first instructed what was then the only GNSS available, the American Global Positioning System (GPS), I detailed specific satellites in orbit, the quality and reliability of their transmission data, their orbital paths, how to minimise the time to first fix and maintain a good fix. Today, other than some very specialist work, such as the use of GNSS at scenes of crime, I no longer teach any of this because the user-friendliness of the technology has advanced so much. This instruction manual dispenses with the vast amount of jargon that initially surrounded GNSS and instead concentrates on the basics of getting to grip with what is a very easy-to-use technology.

The correct use of handheld GNSS makes you safer and more efficient wherever in the world you are travelling and they are affordable, reliable and straightforward to use. I have personally used a handheld GNSS as my primary navigational tool in almost every environment in the world, from the polar regions to the deserts of Africa.

All handheld satnavs:

- work anywhere in the world (there are some limitations at the earth's poles)
- work on any terrain – from magnetised rock, to on the water and in the air
- work in any weather condition with no loss of accuracy

↘ GNSS RULES

→ Learn to navigate competently using a map and compass *first*.

→ Buy the best handheld satnav you can afford.

→ Configure (setup) your receiver before you go out.

→ Learn to use your receiver proficiently in a safe environment before venturing further afield.

→ Always still carry a map and compass, and ideally a spare receiver plus a set of lithium batteries.

- give a very accurate position, displayed as a grid reference, of your exact location and in many models where you are on a local map
- give your height above sea level (elevation)
- can mark and store a current location (called a waypoint or a point of interest) and allow you to navigate directly back to it from anywhere in the world
- can project a waypoint of your choosing and navigate you directly to it from anywhere in the world
- can plot a route and then follow it
- record how long and far you have been travelling (trip time and distance)
- calculate your speed
- calculate the direction you are heading (your bearing)
- provide very accurate local time and date.

Many handheld satnavs also have other inbuilt features which you will learn how to use in this manual, including:

- the ability to record the path you are following
- the ability to track back along the same path
- journey statistics, from distance travelled and average speed to total ascent/descent and estimated time of arrival
- sunrise/sunset, phases of the moon
- alarms to alert you when you are near specific locations
- compass
- barometric altimeter
- camera
- personal location beacon
- the ability to transmit where they are so others can locate you.

A key feature on many receivers is the ability to use digital elevation models to predict upcoming elevation profiles for a planned route.

MOBILE PHONES AND GNSS

There is a revolution happening with mobile phones and it is called satnav!

In 2000 the first mobile phone was launched with GPS capablity, in 2005 there were less than ten models on the market, but by 2011 there were more mobile phones with GPS than all other GPS units ever built!'

The catalyst for this revolution was a piece of United States legislation called E-911, mandated in 2000. This stipulated that when an emergency (911) call is made from a mobile phone, the physical location of the phone must be available to the emergency services within six minutes.

The second force for change was the development of Location Based Services (LBS). These deliver information to the phones user, based on their physical location, such as local weather, nearby restaurants, ATMs and other public service information such as the nearest bus stop or community defibrillator.

GPS was the best globally available technology to meet these demands and was augmented with other positioning technologies, namely mobile phone mast information and Wi-Fi: this is called Assisted GPS (A-GPS).

'I personally use an Apple iPhone both as a mobile phone and as a navigational tool when travelling on foot in urban environments – accordingly, instructions of how to get the most from your GNSS-enabled mobile phone are included in this manual.'

If a GPS signal is poor, or unavailable, as can happen in some urban areas, A-GPS can determine your location using Wi-Fi. If you're not within range of any Wi-Fi, A-GPS can calculate your location using triangulation from mobile phone masts (this last method can be much less accurate, especially in open countryside).

With time the founding technology has continued to improve, with GPS silicon-chips now smaller, using less power and with hugely greater sensitivity – all this requiring smaller antennas: ideal for mobile phones.

Worldwide sales of standalone handheld satnavs are slowly declining, as competition from A-GPS-enabled smartphones takes over. Companies such as Apple, Google and Nokia now essentially provide free navigation with their phones.

Mobile phones & Mapping

Apple have integrated A-GPS into their iPhone, which can be used with Google Earth, Google Maps and various digital mapping programs, where maps can be downloaded and stored so even if there is no available phone signal the GPS still works. Google introduced its own Android operating system for mobile devices, bringing Google Maps capability and navigation to a wide range of mobile phones and Nokia includes Ovi Maps navigation software with its smartphones.

The integration of A-GPS into mobile phones creates much more than a handheld navigation device, it allows you to create a location-based journal where everything is tagged on your trip: photos, videos and even what music you listened (see Nokia's viNe app). In addition, phone users can use LBS to share their location with others and view these locations live on Google Maps, either on their phones or on desktop computers.

Lastly, all of this collective position data can be used to help the emergency services locate individuals with unparalleled speed and accuracy.

Mobile phone apps

An app is simply application software tailored to a mobile phone and they are the key to utilising all of your mobile phones functions. They range from Facebook to the Oxford Dictionary.

Apps that use the A-GPS, such as Google Latitude, are very popular. This app allows a phone user to permit specific people to view their current location on Google Maps, it even includes a feature called 'Location History' which stores and analyses a phone's location over time. Most digital mapping programs, such as Anquet and Memory Map, are available as mobile apps.

Mobile phones – the drawbacks!

Currently there are three limitations with GNSS-enabled mobile phones:

Batteries

The batteries in these devices have to support many functions, including the phone, and the products are aimed at fashionable consumer markets where size matters – therefore large battery packs are not included and battery life for satnav usage can be short. Importantly, the vast majority of mobile phones will not accept standard batteries, either rechargeable or disposable, which can be changed in the field. Continuous power and backup power are a pre-requisite for any crucial piece of equipment.

Durability

Your navigational device needs to be proofed against rain, snow, dry and dusty conditions, extremes in temperature and rugged enough to suffer knocks and drops: most of these phones currently are not (see IP rating, p. 266).

↘ *ULTIMATE NAVIGATION MANUAL* APPS

→ Interactive apps have been especially created for this manual and are designed to help you master the techniques described in this book and are downloadable from **www.harpercollins.co.uk/unmapps** and where we will continue to place apps as and when developed. It is, of course, important to check that an app is compatible with your phone's operating system: Android, BlackBerry OS, Windows Mobile, Symbian or Apple iOS.

All of your eggs in one basket!
If the phone fails, or you lose it, you will have lost your mobile phone and your navigation device. This means that unless you have backup systems you can neither call for help or find your way back.

At present I use an Apple iPhone both as a mobile phone and as a handheld navigational device, mainly in urban environments and occasionally on leisurely walks in the open countryside. I do not use it as the primary navigational tool in either particularly hazardous environments, such as mountains, on extended journeys, or when on a SAR mission – instead I use a standalone satnav, a separate mobile phone and carry a PLB (personal location beacon).

This technology is rapidly changing and more rugged smartphones are being introduced. As this equipment evolves these drawbacks will be minimised, or completely overcome in some instances, definitely in waterpoofing and battery technology: I have no doubts whatsoever that they will become for many, the navigational tool of choice.

Managing the drawbacks

Batteries
To maximise battery life, use the device settings to shut down all non-essential power drains:

1 If you do not need your mobile phone on, turn it off. You can turn it on from time to time and check for messages or make calls.

2 If you are in an area where there is no mobile phone reception, turn the phone off.

3 In open countryside turn off Bluetooth, Wi-Fi and high-speed mobile data services.

4 Set the device screen brightness to a minimum, and disable any 'auto brightness' features.

5 Do not put the phone into **Airplane** mode, as this will power down the phone's subsystem completely and GPS will be disabled.

6 Buy a battery extender, these are batteries, usually rechargeable, that attach externally to your phone, and are relatively cheap.

Durability

1 There are various tough cases you can enclose the phone in: some, such as Magellan's ToughCase, also include a battery pack and are rated to IPx7.

2 Ensure your GPS accuracy is maximised by setting the correct date, time and time zone. If possible, use **Set Automatically**.

3 Incorrect settings on your home computers can sync to your phone – verify the date, time, and time zone settings on any computer that syncs with your phone.

4 Confirm that you have a mobile or Wi-Fi network connection. This allows the A-GPS to locate visible GPS satellites faster, in addition to providing initial location information using the Wi-Fi or mobile networks.

5 Keep a clear view of the sky – structures such as buildings and walls in addition to natural features can block line of sight to the GPS satellites. If this happens, A-GPS will automatically use Wi-Fi or mobile phone networks to determine your position, until the GPS satellites are visible again.

All of your eggs in one basket!
Carry a map and compass, plus a back-up mobile phone.

Using an A-GPS enabled phone to navigate

The following instructions are for the iPhone – most other smartphones function in a similar way.

Urban navigation
In urban environment the iPhone pinpoints your location quickly and accurately using A-GPS and as you move the phone updates your location automatically. The phone displays your location with a blue marker and if your location can't be determined precisely, a blue circle also appears around the marker. The size of the circle depends on how precisely your location can be determined—the smaller the circle, the greater the precision.

At any time you can drop a pin to mark your location (create a waypoint) and share it with others. This function is useful both with friends and family and in SAR situations, the data can be sent via:

- MMS (Multimedia Messaging Service) which is an extended SMS (Short Messaging Service) that allows you to include longer text, graphics, photos, audio clips, video clips, or any combination of these, within certain size limits
- Google Latitude
- Email
- Use the phone to call the information in to another party.

Using LBS

When Location Based Services are enabled and active, a purple pointer appears in the status bar. LBS is on by default, you can turn it off if you want to conserve battery life. You can also individually control which applications have access to it.

For example, if you are searching for an ATM:

❶ Type 'ATM' in the search field within **Maps**. All nearby ATMs will be displayed on the screen, represented by pins. Tap the one you wish to bring up more information about, such as the bank it is in, the bank's phone number, physical and web addresses.

❸ You can search using business names, business types or specific addresses. You can also switch between map view, satellite view and hybrid view and you can double-tap or pinch to zoom in and out on a map.

❹ Choose to view 'walking directions'. The iPhone now gives you a list of turn-by-turn directions, or produces a highlighted map route, and tracks your progress as you move. You can even see what time the next train or bus leaves with public transport directions.

As you view the map on the iPhone's display and your location, the in-built digital compass orients the map automatically to match the direction you are travelling in.

Open countryside navigation

It is essential to download a digital mapping application that accesses topographic maps at 1:50 000, or ideally 1:25 000, scale (1:24 000 in the USA).

In the UK I use Anquet Maps iPhone (the desktop PC version is detailed in the **Digital Mapping** section, see pp. 334–9) and in North America, for wilderness navigation, I use Memory Map or TopoPointUSA by Ebranta Technologies. (TopoPoint includes free United States Geological Survey high-resolution 1:24 000 topographical map images and the app represents great value.) With Memory Map you pay for the maps but the software has more functions and greater flexibility in recording and storing data.

All digital mapping apps have the similar functions to those detailed below.

Using Anquet Maps iPhone

Getting Started

❶ Download Anquet Maps from the Apple App Store onto your iPhone.

❷ On the first run of Anquet Maps, it will ask you for your Anquet login information. If you are a current Anquet user, this is just your email address and password. If you don't yet have an account, you can create one from within the software.

❸ The **Map Manager** will be displayed and will list all the freely available maps plus any maps you have already purchased – you can download these as many times as you need.

❹ Tap on a map to download it.

Moving Around the Map

1 Touch the map and move your finger to move the map.

2 To zoom, just use the standard pinch to zoom as used elsewhere on the iPhone.

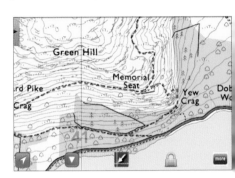

Accessing the Menu and Other Screens

1 Double tap the map, and the menu buttons will appear.

2 Move the map again, and the buttons will disappear, giving you a clearer view of the mapping.

Changing Location

1 Double tap the map to access the menu.

2 Tap the middle button for the Datum List screen.

3 Select either National Grid, Latitude & Longitude or UTM

National Grid

1 If Prefix letters is turned on, this screen will take grid references of the format 'SP 108 368' the amount of precision is determined by the amount of numbers you put for the Easting and Northing within the grid square.

2 If Prefix letters is turned off, this will take an Easting and Northing from the origin of the Ordnance Survey maps. In this case, you will need to set a precision which is easily done on the screen.

Latitude and Longitude

This screen can work in either WGS84 map datum or OS GB. GPS units by default work in WGS84 and Latitudes / Longitudes that are quoted. The screen can take inputs in the following formats:

- Deg.dd Deg.dd
- Deg Min.dd Deg Min.dd
- Deg Min Sec.dd Deg Min Sec.dd
- (N/S) Deg Min Sec.dd (E/W) Deg Min Sec.dd

UTM

This screen can take a fully qualified UTM grid reference of the format 'Zone 31 6922 47964'.

Changing Maps

1 Double tap on the screen to see the menu buttons.

2 Double tap the second button from the left to access the Map Menu.

This screen is split into three areas:

I. The list at the top is the list of map types that you have on your iPhone that are available at the current location. Tap on any of these maps, and your geographical location will not change, but you will be viewing a different map type. This is really useful to quickly switch between maps. Say from 1:25 000 scale mapping to a smaller scale map to get a better feel for the wider area around you then choose again to move back.

II. The second area lists all of the maps that you physically have on your iPhone. These are listed by product, such as Lake District National Park. Tap on the product name, and you will be presented with a list of maps in that product. Tapping on any of these maps, will take you to that map, and hence change your geographical location.

III. If you didn't find the mapping you are looking for, the button at the bottom of the screen will take you to the Map Manager where you can download more mapping.

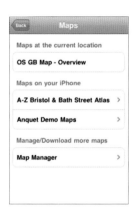

GPS

1 Double tap the map screen to view the menu buttons.

2 Tap the far left button to start the GPS.

3 Crosshairs will appear showing your location.

4 Tap the same button to turn off the GPS.

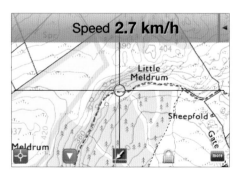

Lock Screen

1 Double tap the map screen to view the menu buttons.

2 Tap the second button from the right – the lock button.

3 This will prevent unwanted interaction with the device whilst in your pocket.

Information Panels

1 At the top of the map screen on the left you will see a grey arrow.

2 Tap this arrow to display the information panel.

3 The panel has 3 panes. To switch panes, just swipe your finger left or right.

4 To hide the information panel, click on the grey arrow on the top right of the screen when the panel is being displayed.

Customising the data display

1 Double tap the map to bring up the menu buttons

2 Tap on the More button

3 Tap on Information Panel

4 Tap on the Panel which you wish to change the data for: left, central or right.

5 Tap on the data field you wish to change, and using the scroll wheel choose the data you wish to view such as location or speed.

The left and central information panels can contain 3 data items, whilst the right panel only displays one data item, but in a much larger format, which allows for easier viewing.

Using TopoPointUSA

Getting Started

1 Open the app and download the USGS maps of your choice directly to your iPhone before departing the mobile data service area or Wi-Fi connection and store them on your iPhone so they can be used at any time.

2 Switch on your A-GPS function.

3 Your phone will automatically download and display map images for your current location and when offline, it will automatically display stored maps in the main screen.

4 Your GPS position and GPS altitude are displayed on the map

5 As you move the map will automatically track in the direction you are travelling.

6 'Swipe' scrolling away from your current location invokes browse mode.

7 Pressing the 'locate' button resumes tracking of your current GPS position. Coordinates are displayed in decimal degrees format.

Detailed Latitude and Longitude coordinates are displayed when browsing map images away from your current location and the software uses the iPhone digital compass to display true or magnetic north. (USGS maps are displayed in true north convention so using magnetic north is not recommended). The iPhone determines heading values and magnetic declination in 'real time' and is not affected the earth's magnetic field 'drift'.

Displaying course, direction and speed

1 Press 'M' (motion) button on the main screen

2 An overlay is displayed containing GPS course, horizontal and vertical speeds

3 The motion button can be hidden by disabling the feature in preferences.

Creating a waypoint

1 Press the single main screen button

2 You current GPS location or browse location is recorded by simply tapping the coordinate display text button in the bottom toolbar.

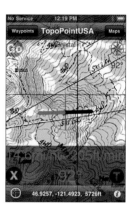

Tracking

1 Enable the 'T' (Track) button

2 Define the recording interval (1–60 minutes).

Browsing

To see an area using the aerial/street view you can jump to a coordinate then you can view can scroll and zoom to any location in the world, just like the iPhone's native map application.

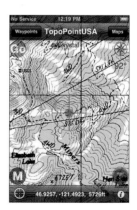

Coordinates

Browsing an aerial/street map view, or searching the USGS feature name database. Latitude and Longitude values are set in the 'Go' screen fields when searching or browsing is complete. Pressing 'Go' will then 'jump' the main screen to the Topographic map of the area.

✕ **QUIRKY FACT:** Apple's iPhone keeps track of where you go, saving the latitude and longitude of the devices' recorded coordinates, along with a timestamp, to a secret file which is then copied to the owner's computer when the two are synchronised – if your phone or computer are stolen, these details can be revealed using straightforward software!

BUYING A GNSS RECEIVER/SATNAV

This manual focuses on handheld satnavs and they are referred to simply as receivers. The same operating procedures can be applied to all personal navigation devices (PNDs) from mobile phones to rugged computer tablets.

Receivers fall into two main categories:

Garmin eTrex H –
a high-quality,
simple handheld
satnav (IP7)with
LCD screen.

Trimble Yuma – a rugged
(IP67) tablet computer
with GNSS and mapping.

- **Basic** – displays your position on the screen as a grid reference (lower price).
- **Mapping** – displays your position on an image of a map (higher price)

Basic receivers are rapidly being superseded by mapping receivers as the technology continues to improve. Millions of basic receivers have been sold and are still in everyday use. Many can be bought cheaply second-hand and are excellent to carry as a backup to your mapping receiver.

If you buy a receiver that's more than three years old, bear in mind that manufacturers often stop providing software/firmware updates. For a basic backup receiver this can be extended up to five years.

Basic or mapping?
Basic receivers
If you prefer to navigate using a map and compass and/or cost is a major consideration these are an excellent addition to augment your navigation. *Throughout this manual these are referred to as* Basic Receivers.

Mapping receivers
These make navigation much more straightforward and mean that your map and compass can stay in your rucksack much more of the time. *Throughout this manual these are referred to as* Mapping Receivers.

GNSS

Selection criteria

There are six key areas to address in selecting which GNSS device suits your needs –
these are:

- robustness
- battery
- screen
- controls
- reliability and backup
- features.

Robustness

Check the ingress protection rating. The IP code is an international classification of
the degrees of protection provided against the intrusion of dust and water and affords
consumers with more detailed information than vague marketing terms such as 'weather
resistant/suitable for outdoors/waterproof'. It consists of the letters IP followed by two
digits (plus an optional letter). This summary will help you choose which one you need.

IP Codes	First digit: Protection against ingress of solid objects	Second digit: Protection against ingress of liquids
0	No protection	No protection
1	N/A	Condensation
2	N/A	A small droplet
3	Solid objects over 2.5 mm e.g. grit	Mist
4	Solid objects over 1.0 mm e.g. sand	A splash from any direction
5	Ingress of dust is not entirely prevented, but will not interfere with the operation of the equipment	Light rain
6	Totally protected against dust ingress	Heavy rain
7	N/A	Submerged in water up to 1 m for 30 mins
8	N/A	Continuous underwater use

If one of the IP digits are replaced with an X this means that the device has not been
rated for this part of the test. IPX7 signifies it is waterproof in up to 1 m of water for
30 mins but has not been rated against solids.

I only use units rated to IPX7. For my iPhone I have bought an external toughened
waterproof case that includes extra battery power.

- **Shock resistance** – is the casing tough enough to withstand dropping it by accident
onto a hard surface?
- **Lanyard** – every essential item should be secured by a lanyard in case of a fall or
accidental drop. Make sure there is an eye/recess where you can securely attach
a lanyard. Many units come with a karabiner and while these are really handy they
should also be used with the lanyard just in case you drop the receiver while holding it.

Battery
If you are navigating in a challenging environment your batteries must be field replaceable.

- A minimum of 12 hours' normal usage: the longer the better
- can use both disposable and rechargeable batteries – especially precharged NiMH
- the receiver can operate using lithium batteries, in cold conditions these batteries are the most reliable and have the longest life – plus they are lightweight!
- can be used/charged from a car-adapter, so if used as a car satnav en route to the start of a walk, battery power is either not used unnecessarily or the rechargeable batteries are charging to full capacity.

For choice and types of batteries see **Batteries** section.

Screen
Size – the larger the screen the larger the area of the map you can see without having to pan out and reduce the scale of the map although this needs to be balanced with a receiver that can still sit in one hand. Screen size is usually given as the diagonal dimension of the visible screen. I recommend mapping receivers with a minimum screen size of 66 mm (2.6 inches). Basic receivers without mapping don't require such large screens.

Resolution – like televisions the higher the resolution the higher the quality of the image and the better the scrolling. Screen resolution is measured in pixels. A screen described as 160 x 240 has 38,400 pixels and should be the minimum you consider.

Sunlight readable technology – these screens do not require the use of a backlight and can clearly be seen in daylight, the major advantage being conservation of your battery power.

Backlight – different natural lighting environments need different levels of backlighting so choose one with an adjustable backlight (note that full backlight exerts a large drain on your battery).

Clarity – most maps use colour and you need to be able to view the map in the format it was created so a colour screen is essential for **Mapping Receivers**. Also you need to be able to clearly see the displayed map in all lights and all weather conditions, *so check it in bright sunlight*. Some receivers have an optional large alphanumeric data which is clear to see.

Touchscreen – good but not essential.

- Make sure you can lock the screen so keystrokes are not entered by accident.
- Choose one which is finger-operated over stylus-operated and avoid capacitive touch technology as they do not work with gloves on.
- Can you customise the data displayed?
- Screen protectors – scratches distort the image and screens are expensive to replace so make sure screen protectors are available.
- 3D – viewing the map as a three-dimensional image of the landscape is a very useful feature to have.

Controls
- Can you use the buttons with gloves on?
- Are the controls clearly marked?
- Can you lock the controls?

Reliability and backup
Read user reviews on the internet and check manufacturer's claims, in particular:

- The software is reliable and not prone to 'freezing' or bugs?
- Software updates are easy to obtain and free: firmware upgrades are important.
- Claims of good satellite acquisition and times to get a fix are accurate?
- The claimed battery life is accurate?
- Which maps are available and what area do they cover?
- Technical backup – does the manufacturer's Helpline have a good reputation?
- Is their website helpful, with FAQs and a Q & A section?

Compatibility
- Is the receiver easy to connect to a computer via a USB or does it require adapters.
- Does the unit work with the digital mapping you own/wish to buy.
- Make sure that the receiver has the Grid System/Map Datum for the area you will be navigating in (i.e. The British Grid 'OSGB,' or the French 'Lambert'). ***This is essential***.
- Wireless connection to other units (Bluetooth or ANT+) is a real benefit for groups that wish to use satnav.

Features/specifications

Must-have features
These are the features I specify when buying a mapping receiver, followed by my 'nice to have' preferences.

Maps
A ***must-have*** for mapping receivers is that the maps display contours – i.e. topographic maps. Many car satnavs only display location without elevation data; however, this is essential for outdoor navigation.

All mapping receivers come with a base map: typically a large-scale world map with little detail, so you need the ability to add more detailed regional maps. Check that the maps you want to use in your region are available for your receiver before buying.

Memory
Minimum internal memory – 500MB. The larger a receiver's memory, the more mapping, trip data (including waypoints)and routes that can be stored. You do not want to have to carry multiple memory cards with additional mapping. If maps are downloaded directly into the receiver's internal memory, the larger the memory the greater the map area you can store. If the maps are available as data cards, the internal memory does not need to be so large.

Waypoints and routes
The more waypoints (also called Points of Interest) and routes your receiver can store the better. The current receiver I am using stores 2,000 waypoints and 200 routes. For basic receivers, storage for 500 waypoints and 20 routes should be the minimum.

Track points
Storage for 10,000 a minimum. This is the breadcrumb trail which your receiver leaves behind you as you move – which you can re-trace using **Trackback**. They can be set to time or distance intervals. If you are navigating tricky terrain and want to record

your movement precisely, storing a track point every second will fill this allowance in only 2 hrs 45 min … a waypoint every 3 seconds in just over 8 hours. Remember that walking at 5 kph you cover 1.4 m of ground every second and over 80 m every minute!

If the receiver is set to automatic recording, it usually stores a point every 10 m of travel, allowing you to store up to ten 10 km tracks.

Electronic compass

Should ideally be three-axis, tilt-compensated. There are two types of compasses built into satnavs:

Electronic compass.

- differential compass – uses satellites to determine the cardinals; does not work when you are stationary.
- fluxgate compass – works irrespective of movement; if your receiver has one of these it will also have a differential compass. There are two types of fluxgate compasses: two-axis, which require you to hold the receiver level for the compass to work correctly; three-axis, which are tilt-compensated and do not have to be held level for the compass to function.

Wireless connectivity

I place a great deal of emphasis on a receiver's ability to transfer information between receivers and with computers easily. Wireless connectivity allows you to share waypoints, routes and tracks while outdoors with other people, and subsequently download this data to digital mapping on your home computer or publish and share them over the internet. Wireless connectivity allows you to seamlessly achieve this without the need for plugs and cables.

There are two principal types of wireless connectivity to consider.

• Wireless Personal Area Networks (WPAN)

A WPAN is a computer network used for communication among computer devices, including mobile phones, personal digital assistants (PDAs) and personal navigation devices (PNDs) in proximity to an individual's body. The devices may or may not belong to the person in question. The reach of a WPAN is typically a few metres. These systems have no subscription fees or ongoing running costs and because the data is transferred over a short distance, who receives it is entirely within your control. As the name suggests they are ideal for personal use.

WPANs use network technologies that include ANT, Bluetooth, IrDA, UWB, ZigBee and Z-Wave. The two most common are ANT and Bluetooth – their advantages/disadvantages are as follows.

ANT – used by Garmin, Nike and Suunto, it transmits up to 10 m and has a low computational drain so uses little battery power. The disadvantage is that most computers do not come with an ANT receiver although you can buy external receivers that connect via USB.

Bluetooth – used by most wireless devices and excellent over short distances. As it uses a radio (broadcast) communications system, devices do not have to be in line of sight of each other. Transmission range varies across three Bluetooth classes:

Class 1, range about 100 m; Class 2, range about 10 m; Class 3, range about 1 m. Most computers support this technology as an inbuilt function.

- **Wide Area Networks (WAN)**

These are designed to transmit data from hundreds of metres to globally, and should be considered by SAR responders and people who will be navigating in remote areas. All these systems have ongoing running costs. A synopsis of types and advantages/disadvantages follows.

3G/GPRS/GSM – these days, because many PDAs/PNDs are mobile phones, the most obvious connectivity is using mobile phone networks. In urban areas these networks have excellent coverage but in more remote areas connectivity can be an issue. You will either pay for the data that you send or will have a subscription fee, sometimes both.

Satellite communications – at one time the use of satellite communications was restricted to government agencies, the military and people who could afford it – it was very expensive. But today satellite constellations, such as the Iridium SC, which is a system of 66 satellites covering the whole earth, including the poles, and is used for worldwide voice and data communication, have become accessible to the general public. Others, such as Orbcomm with 44 satellites in orbit, are being investigated for this type of use. The whole field of satellite communications is opening up and I predict will continue to become yet more readily available at affordable prices to the general public. As both a SAR responder and frequent navigator in remote parts of the world, these are my preferred communication platforms. The battery draw is much more significant that WPANs and the subscriptions fees run for as long as you use the system.

Two-way radios with satnav – I mention these as some of the MRT/SAR teams that I instruct have them. This technology combines a satnav with a communications radio to provide realtime tracking. They are restricted by line of sight on many frequencies and inferior to both 3G/GPRS/GSM and satellite communications.

Other technologies – worth mentioning are TETRA (Terrestrial Trunked Radio), used in the UK by most emergency services and the intelligence services (a highly secure encrypted system with real-time location easily provided). Similar to mobile phone connectivity, use is restricted by the proximity of the transceiver to a TETRA mast which, in remote areas, can be low if at all. APRS (Automatic Position/Packet Reporting System) is a technology based on amateur radio frequencies; very cheap to run but again limited by the proximity of communication masts.

Barometric altimeter

All satnavs will give your GNSS elevation if they have reception of four or more satellites (called a 3D fix) and are accurate to on average three times the stated horizontal accuracy. Having a barometric altimeter gives you the option of checking the GNSS elevation figure – some receivers do this automatically and are accurate to 1.4 times the horizontal accuracy. (Some satnavs can also provide the elevation from the digital maps.)

Automatic routing

Turn-by-turn routing on roads allows the receiver to be used as a car satnav when you are driving to your start. Often the detailed road mapping is an optional extra. Automating Routing is also being extended beyond roads to include trails.

Custom-map compatible
This is a feature which allows you to download any electronic map, or scanned paper map, *even one you create*, onto your mapping receiver – from a satellite image of your home area to a hand-drawn map of an incident in MR.

Waypoint averaging
Allows you to refine a waypoint location with multiple samples to achieve the most accurate location possible. This is a must-have for MRT/SAR teams and other agencies attending incidents/accidents.

Proximity alarms
A proximity alarm allows for the marking of waypoints which have an invisible radius around them – if you enter this space an alarm sounds on your receiver. They are excellent to use for designating areas of danger, particularly those not marked on the map – such as small potholes, or a sudden steep drop. Like waypoint averaging this is a must have for MR/SAR.

Sun and moon tables
These detail the exact time the sun will rise and set on a particular day and therefore allow you to calculate daylight hours to travel. The phases of the moon can also be detailed for specific days and allow you to determine possible available moonlight to navigate by.

Nice-to-have features

Camera
Photographs can be geotagged (each photo is location/time/date-stamped) which is an additional benefit for recreational use. In SAR it provides an excellent tool for recording the incident locus and as such would move up to the must-have list above. See also **Back Snaps** (p. 141)

Remote monitoring
This feature allows your (the receiver's) location to be viewed remotely online. Like the camera, an additional benefit for recreational use but a serious contender for must-have in SAR.

Voice recorder
The ability to record speech and geotag the recording. The same priority as a camera or remote monitoring.

Personal Location Beacon – PLB
I personally would not have my primary navigational tool and PLB in the same device – if the receiver fails, you instantly lose not one, but two vital tools.

Short Messaging Service – SMS
Not having to rely upon a mobile telephone network, instead using something like the Iridium global network of satellites to send/receive text message is a safety feature for the general public and an important tool for SAR *if cost-effective*.

Phone
An inbuilt mobile phone. Again I personally would not have my primary navigational tool and mobile phone in the same device because if the receiver fails, you instantly loose two vital tools; also the significant additional battery drain reduces the time you can use the device before replacing/recharging the batteries.

Future features!
At the time of writing I have used some technologies still in development – if they become available they should be included in your checklist and prioritized according to your needs and budget!

- **Enhanced Positioning technology** – the satnav uses movement and gravity sensors to calculate your position when you temporarily lose satellite reception.
- **Synthetic Vision technology** – uses sophisticated graphics modelling to recreate a visual topographic landscape. As you hold the unit out in front of you a 3D map image is shown of what is in front of you and as you rotate, so does the image. This will be a massive advantage in low-visibility conditions such as bad weather.
- **Location Reporting** – where you can be seen by others and you can see others. Receivers have the ability to communicate using phone SIM cards 3G/GPRS/GSM/SMS or Satellite Communication Systems such as the Iridium Network and radio broadcasts TETRA (*Terrestrial Trunked Radio*) or APRS (Automatic Position/Packet Reporting System based on amateur radio frequencies).
- **Data Transmission** – using these communication technologies everything from geotagged photographs can be sent to text messages.

BATTERIES

Battery technology has made massive strides alongside the advances in mobile phones and handheld navigation devices and the array of different types can be quite bewildering.

You have probably spent a lot of money buying the best handheld satnav and head torch for your purposes – do not skimp on cheaper, inferior batteries. If your devices have no power they are of no use whatsoever.

I have used many different types of battery in environments ranging from crossing Lake Fryxell in the Antarctic, to the extreme heat of the Makgadikgadi Pans in the Kalahari Desert. In addition, I have spent years dealing directly with both the manufacturers of batteries and satnavs and this section cuts through the technical jargon and tells you exactly what to look for, so they never let you down when you are navigating.

Types of battery

- Lithium-ion batteries – (sometimes also called lithium polymer) are the type supplied with nearly all mobile phones and handheld satnavs by the manufacturer, who will have gone to a lot of time and effort to produce a battery which is specific to their device's needs. They are the everyday battery of choice for the unit and are rechargeable and lightweight and work well in extreme temperatures.
- Rechargeable – are the batteries for everyday use if your unit does not have a lithium-ion battery. They are available in two types, standard rechargeable and precharged reusable batteries. While they both have an initial higher cost than disposables, standard can be recharged hundreds of times and precharge more than 1,000 times, if looked after properly, so are very cost effective – plus they are both more environmentally friendly. The precharged are superior to the standard rechargeable as they can hold their charge, if unused (95% after one year and 75% after three years). Plus they can operate at lower temperatures than standard rechargeable batteries.

- Disposable batteries – are the perfect backup system and essential if you do not have access to a battery recharger. The two main types are alkaline and the more expensive lithium (not to be confused with lithium-ion batteries) and can have a storage life of 7 and 15 years respectively.

Jargon Busting

mAh – this is the capacity of the battery, the higher the mAh the more run-time you will get from your device.

NiCad – nickel cadmium batteries have become much less popular primarily because they contain the highly toxic metal cadmium and are difficult to dispose of without causing pollution. Also battery life can easily be shortened if NiCad batteries are not fully discharged between charges, which is a real hassle – I never use them.

NiMH – nickel metal mydride are the best type of rechargeable battery as you can recharge them at any time, whether discharged or not, and they have high capacities (mAh). The superior, precharged reusable NiMH batteries work well down to –10° C.

Lithium – these are disposable batteries and different from lithium-ion batteries. They are fairly expensive; however I always carry a set as a backup because they have a storage life of 15 years, can be used in extreme temperature conditions (between –40° C and 60° C) and are lightweight. They have the highest capacity (mAh) of all the batteries, giving you the longest run-time.

Alkaline – are a type of disposable battery or rechargeable battery. They are relatively cheap and have a good capacity and storage life of up to seven years. However, they operate best at room temperature with performance dropping off rapidly at lower temperatures: an alkaline battery producing 250 mAh at 0° C is only half as efficient as it is at 20° C.

Obviously battery life will vary from manufacturer to manufacturer, receiver to receiver and user to user and in different environments. To give you an idea of the approximate amount of user time you can expect from the different batteries per charge or single-use disposable batteries I have recorded my own use over hundreds of journeys and these are my findings. The lithium disposables give the longest power so count as 100%, all other batteries are a percentage of this. So if you get 24 hours' use with a lithium disposable I would expect to get six hours with a standard NiMH battery.

- Lithium disposable   100%
- Lithium-ion (when new)   90%
- Alkaline   55%
- NiMH   40%

Which type of battery?

The most important factors to consider in choosing the right battery for your journey are: what will be the likely operating temperature and how long will you be away?

I carry two handheld satnavs. My main unit is a modern mapping receiver which I use with its manufacturer-supplied lithium-ion battery. The other is a basic model which I take as a backup, with a pair of lithium disposable batteries, stored dry and separately.

If I am intending to head out on a long trip over many days, I use precharged reusable NiMH rechargeable batteries and carry the full amount of spare, fully charged ones I estimate I will need for my trip.

If my trip is planned for more than a few days (but rather weeks or months) I carry disposable alkaline batteries or, if the temperature is going to be extreme, lithium disposable batteries. You could also take a portable solar recharger and NiMH but you must be certain of both sunshine and the time to capture it.

Battery Rechargers

Buy a charger from the same manufacturer as your rechargeable batteries if you can. Recommended features are:

- ability to use both mains electricity and solar power, and have a USB power port
- automatic cut-out to avoid overcharging the batteries
- status indicator (usually an LED light) which indicates the condition of the battery – discharged/charging/fully charged ... or no longer serviceable
- maintenance charge setting to keep batteries fresh
- temperature shutoff to prevent overheating of cell
- if you intend taking it on trips with you, try to get one which is weatherproof (or keep it in a watertight bag).

Getting the most from your batteries

By reducing the demand made upon on batteries by your equipment, their life can be significantly extended and accordingly the following changes should be made to your handheld satnav where possible.

- Adjust the timeout for the backlight to less than 30 seconds I find 15 ample.
- Use 'Battery Save' mode or the 'hot standby' feature which causes the GNSS to update its position less often in terrain where it is safe to do so. The unit will automatically join up your track points and allow you to quickly obtain a GNSS position to guide you on your way when you need it, just as you would normally use a map and compass.
- Disable the unit when stopped for lunch or moving indoors.
- In cold climates, keep the unit warm by storing in a coat pocket – batteries use more energy when cold.
- Remove the batteries if not in use for more than a month.
- When **Tracking** change the record interval to suit the terrain: on a narrow mountain ridge set it to 'Most Often' and on a large plain set it to 'Least Often'.
- Set the unit to switch off if external power is lost, so that when you disconnect from your computer the receiver will automatically shut down.

Using your batteries on your trip

Make sure that either your batteries are fully recharged or you have plenty of power remaining in your disposable batteries. Do this at the ambient temperature in which you will be navigating – batteries warm from cars or pockets can show more bars on the battery life than there actually is. In practice, when I get out of a vehicle, I place my unit on the roof while I prepare for my trip – when the unit has cooled to ambient temperature, I check the battery strength.

↘ EXPERT TIPS

→ Try to choose equipment that uses the same battery size so you do not need to carry different spares – also, batteries that are interchangeable if needed.

→ To preserve mobile battery phone life, lower the speaker volume and switch the ringer to silent whenever you can. Reduce or disable screen brightness, backlights and displays. Use 'power save' settings where possible and if you can, turn the phone off for a while.

→ If you store rechargeable lithium-ion batteries in a fridge they can hold their charge for a year.

→ Lithium-ion rechargeable batteries, even if kept in perfect conditions, will permanently lose as much as 20% of their maximum charge capacity per year due to internal oxidation. If exposed to high temperatures or if kept at full charge for extended periods, this deterioration in performance can be even faster.

If working in extremely cold conditions, keep your unit warm inside your jacket and only take it out when you need to confirm location. Remember to hold it correctly and allow it to gain full satellite acquisition.

If you need to change batteries, only open the unit if it is sheltered, even if just inside your jacket, and remove the old batteries and put them somewhere separate before you install the fresh batteries, so you do not accidently mix them up.

Battery maintenance

All batteries lose their charge when not in use. The rate of loss varies by type. The worst are standard NiMH, which can lose 0.5–1.5% per day, and the minimum loss is 30% over six months. The best disposable are lithium, which retain the majority of their charge for up to 15 years.

NiMH rechargeable batteries do not have the memory problems typically associated with older nickel cadmium rechargeable battery technology, so they can be recharged after a few hours of use or whenever convenient.

- Store batteries in a dry and cool environment.
- Never expose any batteries, especially lithium, to direct sunlight.
- Remember that all rechargeable batteries have a life and will eventually need replacing.
- Never mix different battery chemistries as they discharge differently in a device, which could potentially lead to premature battery failure of the lower-capacity batteries. Avoid mixing different makes of battery for the same reasons.
- Do not mix charged and discharged batteries in the same unit – this can cause something called reverse polarity which damages the batteries.

YOUR NEW SATNAV CHECKLIST

As technology continues to develop, many of the following actions will no doubt become automatic and be seamlessly carried out in the background without you needing to do anything. Notwithstanding this, some, such as applying screen protectors, are physical tasks. Others, such as checking the stated accuracy need your input. Understanding how your unit configures itself will help you troubleshoot when something goes awry.

Even if you have already been using your unit for some time, this section is a good 'Back to Basics' to ensure that you have missed nothing in its set-up, even small omissions, such as the incorrect map datum, can severely impair the correct functioning of your unit.

All manufacturers provide some sort of a quick start guide to get you up and running; take time to go through it with your unit. The 20 mins it will take you to run through this checklist will save you hours of frustration in the future! This is a general checklist for all handheld satnavs; if your unit does not have the function referred to or this step has already been configured skip it and move on to the next.

Protect

	Action	×
1	Remove the thin film of plastic protecting the screen and replace with either a screen protector to guard against scratches or buy a case for it.	
2	Attach the paracord lanyard 0.75–1 m (30–40 in) in length, even if it comes supplied with a karabiner. The lanyard needs to be long enough so you can hold your receiver with your arm outstretched, and short enough not to hit the ground if dropped.	

Prepare

	Action	×
3	Open battery cover and check that polarity is clearly marked. You need to be able to see this in poor light levels or weather. If it is not (most units are not) affix self-adhesive colour stickers over the '+' signs. 	

4	Select which type of battery you are going to use (**Batteries section**).	
5	Insert batteries that have a full charge observing correct polarity.	×
6	Turn the unit on.	
7	Select the language and go to Setup.	
8	Set the correct battery type – this is very important to get the maximum life from your batteries.	
9	Enter contact details, key in 'If found please call' and give your mobile telephone number only not your home address. On some satnav you can enter these details into a file, usually called 'startup.txt', which you access when the unit is connected to your computer.	
10	Enable SBAS (such as WAAS/EGNOS) if not already enabled	
11	Select Position Format.	
12	Select Map Datum. On many devices this is automatic when you set up the position format (see **11**).	
13	Select measurement units – choose between metric or imperial: opt for the same units as the maps you will be using.	
14	Set heading to degrees.	
15	Set the clock display: MR & SAR select '24 Hrs'	
16	Change backlight to lowest level comfortable to use – this is the biggest potential drain on your batteries.	
17	Turn off key stroke beeps.	
18	Set up map.	
19	Orient map to Track Up this is the same as orienting a printed map (see p. 93). If it is set to 'North Up' you will view the map display as you would if holding a printed map upright.	
20	Select either road routing or off-road depending upon environment.	
21	Make a note of your grid reference either from a map, from Google Earth or another digital mapping application. Make sure you use the same map datum!	

Position

	Action	×
24	Acquire full satellite signal by going outside where you have a clear 360° view of the sky, or as near as possible, free from trees and large buildings. This can take up to 5 mins.	
25	The unit will either give your position in lat/long (usual default) or use your local grid system. Compare this reading with the one you recorded from your map/Google Earth. As long as they are close to this reading this is fine at this stage.	

26	Calibrate the compass, if it has one. Do not stand near objects which interfere with magnetic fields such as cars, overhead power lines and buildings.	

Connectivity

	Action	×
27	If your unit came with a computer disc insert it and follow the onscreen instructions.	
28	Alternatively connect your unit to your computer either via the cable supplied or wirelessly.	
29	Go to manufacturer's website and find Support/Downloads (sometimes tabbed as Software).	
30	Follow the online instructions to update your firmware/software.	

Checklist available as PDF at **micronavigation.com**.

Check 7

Check 11

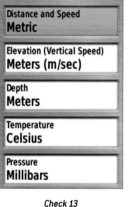

Check 13

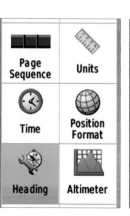

Check 14

Check 14

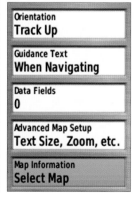

Check 18

↘ EXPERT TIPS

→ Calibrate the compass (rotate the unit, not yourself!):

- After every battery change
- If you move more than 160 km (100 miles)
- The temperature changes ±20° C/68° F
- If you are entering an area where position is critical.

→ Update your software/firmware monthly. Always have charged batteries in the unit and never disconnect the it whilst it is updating.

BASICS SECTION

GETTING YOU STARTED

If you are a competent navigator with map and compass, but have never used a handheld GNSS receiver before, the Basics Section aims to take you to a level of competency whereby you should be able to navigate major routes with confidence.

The same principles of operation apply to all receivers, no matter what make, and similar to cars, the one you learn to drive in may not be the one you keep for life so even as these receivers become out of date the new ones are sure to have the same important basic functions.

I use a principal mapping receiver and as a backup I carry a five-year-old second-hand basic receiver I bought for very little on an internet auction site.

GNSS jargon buster

Waypoint – is a location which you record and store on your satnav. You can navigate to a waypoint from anywhere in the world, but remember that you are only led back to a waypoint in a straight line – your receiver will not take into account natural hazards between you and the objective!

Route – a series of waypoints that are joined together. These can be marked in transit or by plotting and entering them beforehand if your unit has mapping capability. Alternatively, you can create routes using mapping software (such as Anquet) and then download them to your unit before departure. You can even share them with other people via email to download onto their GNSS receiver! Data fields can be selected to provide distance, time en route, arrival time or the name of the next point or destination.

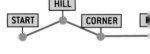

Track – a 'breadcrumb' trail, a record of exactly where you have walked. A track log contains hundreds of track points – not to be confused with waypoints. In the past a track was either simply followed back (Trackback), or converted to 'smoothed', less accurate routes. Improvements in the technology mean that automatic detection of elevation can be stored in the track along with significant landmarks (such a car or basecamp) that are saved as waypoints and added to the map and listed on the active route page. The addition of these waypoints, along with elevation data, creates tracks which much more resemble detailed routes. They can

so display the same data fields as routes. Plus you can plot them directly onto Google Earth or Microsoft Virtual Earth. You can also export them to other receivers or computers as GPX files. A GPX file is a device-independent data format used for GNSS navigation devices – it can describe a route and can be uploaded to most makes of receiver.

location – your location can either be displayed on electronic mapping or described as a grid reference. All receivers calculate your position and display it usually in latitude and longitude. Forgetting to change your receiver's display to the local coordinate or grid system is an easy and common mistake, so, for example, when in the UK change the settings to: Position format > British Grid • Map datum > Ordnance Survey GB • Map spheroid > Airy.

Take time to understand these terms; it will make your learning both quicker and more rewarding!

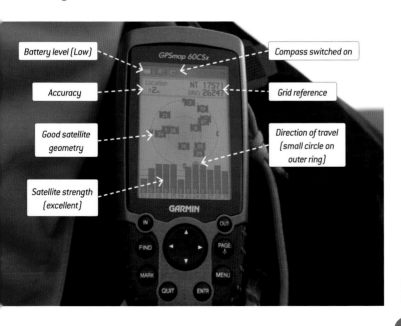

Battery level (Low)

Accuracy

Good satellite geometry

Satellite strength (excellent)

GPSmap 60CSx

Location
±2m   NT 17571
BNG 26247

GARMIN

IN   OUT
FIND   PAGE
MARK   MENU
QUIT   ENTR

Compass switched on

Grid reference

Direction of travel (small circle on outer ring)

How to hold a satnav

Your receiver needs to obtain clear, uninterrupted signals from as many satellites as it can to be accurate. Objects such a mountains or buildings can interfere with these signals and reduce the precision of your device, so learning how to always hold your receiver is very important. Find out what sort of antenna (aerial) your unit has because there are three principal types which operate best in different positions.

- **Quadrifilar helix antenna**: these are easy to identify as they are housed in a protrusion from the main body of the receiver.
- **Ceramic patch antenna**: there will be extra space at the top of the receiver – called a 'forehead'.

Quadrifilar helix

Ceramic patch

- **Linear antenna**: if it doesn't look like there is room for an antenna it's a good guess the receiver has a linear antenna.

The optimal positions to hold these different types of receivers are:

- quadrifilar helix antenna vertical to 45°
- ceramic patch antenna horizontal to 45°

Holding in the optimal position: (left) quad helix and (right) ceramic patch antennas.

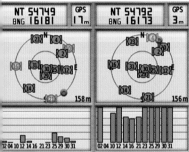

eceiver held horizontally and close to the body eft) – suboptimal antenna position. Receiver held ertically and away from body (right) – optimal ntenna position. The histogram (green bar chart) hows the relative strength of each satellite signal eceived, the higher the bar the better the signal. In the op right of the screen, the GNSS accuracy is displayed nd this is directly related to signal strength, umber of satellites and where they are in the sky.

- linear antennas are more omnidirectional than their counterparts, so position is less important than quad helix or patch.

The data is sent using microwaves and these have difficulty passing through rocks, buildings, trees and most of all you – because the water inside us absorbs the energy very efficiently; therefore hold your unit away from your body and at shoulder height so its antenna has a clear view of the sky.

Look at the satellite screen and check the stated accuracy – if this is improving (the number getting smaller) then wait until it settles.

On the same screen look at the satellite geometry, which in simple terms is how well spread around you hey are. If they are all clustered above you, or to one side, the GNSS accuracy will be ow. If you can see a satellite on your screen which would change this and your receiver s not obtaining a good signal from it (either the strength bar is low or empty) it may be bscured by a tree or other obstacle so move your position slightly. Sometimes rotating he position you are facing by 180° has an effect.

Multipathing

n environments where there are large reflective surfaces which the satellite signals an bounce off, such as buildings or rock faces, your receiver can suffer from multipathing and give an inaccurate reading. Usually your unit's accuracy will fluctuate unpredictably or the compass jump around. If this happens, or just common sense tells you that you are immediately next to such a surface, shield your unit from the stray ransmission with your body:

1 Hold the unit in cupped hands to stop stray signals from any rock that may be beneath beneath you.

2 Face away from the surface you suspect is causing the multipath.

3 Wait for the unit to settle and then obtain you location.

Many advances are taking place to detect, mitigate and correct the impact of multipath and further improvements are likely as hybrid satnavs (employing more than one satellite system) become available.

Alternatively, as GNSS satellites are always moving relative to your position, you can always wait for a better configuration. GNSS satellites will traverse the sky you can see within a 4–6 hour timeframe.

Accuracy

Satnav accuracy varies depending on the quality of signals they receive and where the satellites are positioned in the sky. In canyons they will generally be less accurate than when you have a completely clear view of the sky to the horizon. Receivers do not position you with pinpoint accuracy – instead they place you somewhere inside a circle with varying degrees of probability.

Accuracy is usually displayed on your map page as a circle around the reported location – this circle represents an area where there is a 95% probability of you being.

Accuracy is also shown as a number (usually on the satellite screen). This number is the length of the radius of the 95% probability circle.

What about the remaining 5% chance of you being somewhere else? For 1.5% of the time you will be somewhere within a circle 2.55 times this radius; and there is a 0.1% chance of you being in a circle 5 times the stated radius.

However, in practice you are invariably within the stated accuracy of your exact location, often almost exactly on it!

For a more detailed explanation see the **Advanced Section**.

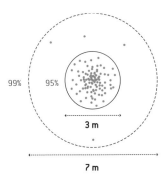

99%   95%

3 m

7 m

Reported location is on the edge of circle – actual location could be on opposite side

Taking the reported location as a pivot, actual location can rotate around this point – this is why we double the reported radius

3 m

6 m

You are somewhere inside this circle of stated accuracy.

earching for a previously created vital waypoint

√hen we are searching for a waypoint previously marked by a receiver these degrees f error can be compounded.

As you move you are somewhere inside a circle of accuracy that moves with you. he waypoint you are approaching was created at a different time with a different atellite configuration and is somewhere inside its own circle of accuracy. Let's say that /hen the waypoint was created the receiver's stated accuracy was 5 m – so is inside 12.5 m diameter circle. Let's also say that your receiver's stated accuracy is now 4 m you are therefore somewhere inside a 10 m diameter circle. There is a possibility that 'hen your receiver tells you that you have arrived, you could be as much as 22.5 m

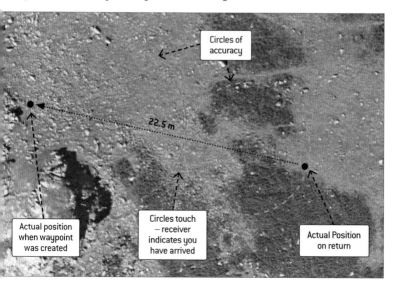

Circles of accuracy

22.5 m

Actual position when waypoint was created

Circles touch – receiver indicates you have arrived

Actual Position on return

from the actual point because they are different circles and you and the waypoint could be on opposing sides of these circles.

In view of this it is important to be continually aware of the accuracy of your receiver when you are creating critical waypoints and navigating to them. You can do little to affect the accuracy when navigating towards this waypoint, other than the best practice advocated in the manual – conversely you have yet much greater control when you are actully creating the waypoint. See **Creating Accurate Waypoints**

Distance accuracy

Your receiver calculates distance from a mathematical estimated ellipsoid sphere (usually the WGS84). This numerical model is stored internally on the unit and does not take account of hills, valley, mountains – the topography. So on the flat your receiver is accurate to its stated degree – however, when climbing or descending steep ground it will be inaccurate! The steeper the incline, the greater the inaccuracy. This uncertainty is compounded by multipathing and the sky above you being shielded from satellites.

With the advent of sophisticated topographic digital mapping in receivers this problem is being tackled and I suspect will eventually be overcome.

Make sure that your satnav is at its optimum antenna position at all times for the greatest accuracy recording your track.

WORKING WITH WAYPOINTS

You can mark any location in the world by creating a waypoint.

Waypoints are at the very heart of using GNSS and while easy to use, you should consider the methods presented here as best practice.

1 Select the satellite screen. Hold the receiver at arm's length and head height in the correct position for your type of antenna position.

2 Wait for about 10 seconds for your unit to settle, especially if you have taken it out of your pocket or it was in sleep mode.

3 If you see the unit's accuracy continuing to improve, either by looking at the stated GNSS accuracy or a bar chart of satellite strength, wait until you feel it has reached its optimum, usually around 30 seconds.

4 Check their geometry: the further spaced to the horizon the better your unit can calculate your position. If they are all grouped above you or to one side decide how important your waypoint accuracy is.

GNSS

Good geometry

Bad geor

5 Mark your location and you will now be prompted to name it. Do not simply use the default number offered because after a few different numbers have been entered you will forget what each waypoint relates to.

↘ EXPERT TIP

→ Use the symbols provided in your units and if necessary add a number to them so if on your journey you mark several gates in fences simply list them sequentially 'Pic-1', 'Pic-2' and so forth.

Navigating to a waypoint

As noted before, a GNSS will take you in a straight line back to a waypoint and take no account of mountains, canyons, rivers, oceans or any topographical feature whatsoever! This is why you need mapping to navigate safely with a GNSS, either a printed map or a digital one installed onto your satnav. All GNSS devices contain an approximation of the world's shape (an ellipsoid sphere) so if you are in London and wish to navigate to a waypoint in San Francisco your unit will direct you in a straight line over the surface of the earth to it – effectively by a great circle route.

Digital trail networks available in electronic cartography have started to use auto-routing which takes into consideration natural features and hazards – but it is still best to think of travelling to waypoints like being hauled in on the end of a fishing line.

As you navigate towards your waypoint, especially if it is critical, keep checking the stated GNSS accuracy – as you near the waypoint your receiver may indicate you have arrived when in reality you may still have a way to go, due to the probabilistic nature of position fixing of GNSS, as explained above (pp. 286–7)

GNSS

→ When you are only a few metres away from your waypoint – dependent upon size – look up and not at the screen because if the item you waypointed is small, say a set of keys, your unit will only usually put you within a few metres of them.

→ When you have reached your waypoint, remember to instruct your receiver to stop navigating to it, otherwise it will keep taking you there when you move away!

Editing a waypoint

When waypoints have been created you can go back and edit them – their name, the symbols used to represent them, even their grid references.

Deleting a waypoint

If you do not intend to use a waypoint or created it in error, it is good housekeeping to remove them.

Projecting a waypoint

A projected waypoint can be anywhere and any distance from where you actually are. This technique can also be called 'forward projecting a waypoint'. A projected waypoint can come from:

> 'This is the most-used satnav feature, and I teach it to all SAR teams.'

- a paper map: finding the point you wish to go to and taking its grid reference
- a digital map on which you have created a waypoint that is then downloaded to your handheld unit
- a person givng you a grid reference either verbally, by texting it, or sending you a waypoint they have created on their handheld GNSS.

When this waypoint is shared with other people you create a rendezvous: for example, a car park where you can all meet up before embarking upon your walk or the location of an accident for other SAR members to attend. There are five ways to do this using your handheld GNNS unit:

- Sight 'N Go
- entering the grid reference
- using the map cursor
- share wirelessly
  from your computer.

Sight 'N Go

I like this feature which is installed on many Garmin receivers. It provides a very quick and simple way to project a waypoint and navigate to a feature within your sight – by simply pointing your satnav directly at it.

Either select Sight 'N Go from the main menu or the compass screen, point the receiver at the feature and lock the direction.

GNSS

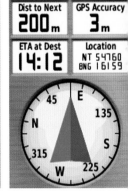

Projecting a waypoint using Sight 'N Go takes just a few straightforward steps.

1 Bring the satnav up to eye level

2 Look across the level unit, like you would sight a gun

3 For units that have a white notch at the top of the screen that you use as the foresight and line up with the white mark on the top of the joystick to aim at the feature.

4 For unit that use the top white diamond on the joystick, use the Quit/Enter button as the rear sights.

5 When you press enter to lock the direction be careful in doing so you don't move the unit.

6 You will be given two choices:

- **Set Course** navigates you in this direction
- **Project Waypoint** allows you to estimate your distance from the feature. Do this using **Stereoscopic Ranging**. You may wish to do this if, in navigating to it, you think you may lose sight of the feature.

Entering the grid reference

If you are given a grid reference, or take one from a map, there are two methods to input it into your satnav.

1 Select **Find**, choose the Coordinates screen and simply enter the location (Grid reference).

2 Create a waypoint in the usual way and before you save it, change its grid reference (in the same way you can change its name) and simply save it.

A note about different-sized grid references: if you have visually taken a grid reference from a map it will probably be a six-figure grid – and at best an eight-figure. Your satnav uses ten- (some use twelve-) figure grid references – to convert your reference to either a ten- or twelve-figure grid reference simply add zeros after the easting accordingly and do the same for the northing.

- Six-figure grid: NT 782598 (easting is 782 and northing is 598)
- Ten-figure grid: NT 7820059800
- Twelve-figure grid: NT 782000598000

- Eight-figure grid: NT 78245981 (easting is 7842 and northing is 5981)
- Ten-figure grid: NT 7824059810
- Twelve-figure grid: NT 782400598100

Using the map cursor

Simply move your unit's cursor to where you wish to create the new waypoint and press Enter when it is over the location you wish to mark.

Share Wirelessly

Some units can send or receive data when connected to another compatible device. You must be within the transmission range of the communications technology (normally 10–15 m) and have a clear line of sight to the other satnav. If your device only transmits to one other satnav at a time, and you are in a group, agree a protocol, such as, 'I am going to send data to named person' and the person receiving replies 'named person received data'.

1 From the main menu, select **Share Wirelessly**.

2 Select Send or Receive.

3 Follow the on-screen instructions.

From your computer

By connecting your device to a computer you can download waypoints you have received via email or waypoints that you have created with your digital mapping software. See **Digital Mapping** section, pp. 329–39.

Man overboard

This is an instant way to create a waypoint of where you are. Used by mariners at sea it can also be used on land: for instance, if a member of your party falls while descending a slope on snow or ice, you can immediately mark their last known position.

It is a ***one-button action*** and immediately after it has been created, the receiver will tell you how far you are away from the waypoint and direct you back to it.

WORKING WITH ROUTES

A route is just a series of waypoints joined together and you navigate from one waypoint to the next in a straight line.

Routes can be created in three ways:

- marking as you go
- projecting waypoints on your satnav and creating a new route
- using mapping software and then downloading onto your satnav.

You can share your routes with other people and also use routes they have created. There are now thousands of websites where like-minded groups, from hikers to mountaineers, can share their routes and waypoints.

Creating a route
From the Routes menu the most straightforward way to create a route on a mapping receiver is to scroll the cursor over the map and plot each waypoint. You can also use waypoints you have already recorded.

Editing the name of a route
Always name your routes – as you build up a collection it is far easier to search for them by name.

Editing a route
The waypoints in a route can be inserted, moved or deleted – this comes in handy if, when you actually navigate the route, you find a change in the path (such as a locked gate), or that there are too many waypoints and you can safely navigate with fewer.

Viewing a route on the map
Just as you would on a paper map, reviewing the context of your route is essential to ensure that you have not made an obvious mistake – such as two waypoints joining over a stretch of water.

Deleting a route
Housekeeping (file management) is important. If you build up a large number of routes over time, your receiver's memory will become full – in which case you will either have to move them to your PC or digital mapping, or delete them.

Navigating a route
Find the route you wish to navigate in the Routes menu and then simply go to it. You can specify if you wish to commence at the beginning or end of the route – some units will default to the waypoint nearest to your present position.

Viewing the active route
You can either view the route you are following on the map or display the bearing (course pointer) of your device.

Stopping navigation of a route
When you complete your route, or decide that you no longer wish to follow a specified route, cancel/stop the receiver from navigating to it, otherwise it will keep trying to direct you back on to it.

Navigating a reverse route
From the route planning menu select the route, then select reverse route and go.

Reviewing a route
Study the length of a selected route, including the calculated total ascent/descent – this will allow you to decide whether it is appropriate for the party/weather/daylight hours.

WORKING WITH TRACKS

Your receiver records a track log while you are moving.

You can save these tracks and navigate them later – many receivers will auto-archive these tracks. Also, you can share them through various websites. If you want to create tracks with greater control and a finer resolution, there are proprietary software packages available, such as Garmin's BaseCamp.

Viewing archived or loaded tracks
Select the Track Manager or Tracks option from your menu.

Managing track log recording
You need to choose the way your receiver records the tracks you create – the options are usually:

- **Distance** – records tracks at a specified distance interval between each track point created
- **Time** – records tracks at a specified time interval between each track point created
- **Auto** – records tracks at variable rate to create an optimal record of your tracks
- **Recording Interval** – an option to record tracks more or less often.

Note: using the **Most Often** interval provides the greatest track detail, but fills your handset's memory rapidly. The maximum number of track points that can currently be stored in a handheld receiver's memory is 10,000 – older points are archived and can be reviewed at any time.

Viewing the current track
You can view the current track on the map – some receivers also allow you to detail the elevation plot for the track.

Saving the current track
Unless your device has an autosave feature, you will need to save your track after you have created it. Some units allow you to save a portion of the track.

Clearing the current track
You only need to do this if your receiver does not provide automatic archiving.

Deleting a track
Memory management again – be aware of available remaining memory.

Navigating a saved track
You must record and save a track before you can navigate it. In your tracks menu select the track you wish to navigate and **Go**.

Archived tracks
One of the great features of tracks you have created and saved is that you can pull them into your favourites list and then easily access them to renavigate anytime.

Track elevation profile
This is useful to determine what your total ascent/descent was during the journey, the distance you travelled and area encompassed.

USING THE COMPASS

Like a magnetic compass your receiver's compass displays the cardinals in reference to your location.

This is the screen that will show you the way to walk when you are following a route or track, or navigating to a waypoint or point of interest.

The bearing pointer

The bearing pointer is an arrow and it always points in the direction of your destination, no matter which way you move.

When the arrow points towards the top of the receiver you are travelling straight towards your destination. When it points in any other direction turn towards this direction until the arrow is pointing to the top again.

On the compass page from the menu select **Data Fields** — some models allow you to select and display only one, while others allow you to select up to four. The information I choose to display in order of preference is:

- **Dist to Next** — how far you have to travel to reach your next waypoint/point of interest
- **ETA to Next** — travel time to next waypoint at your current speed, which, depending upon your model, may or may not take account of changes in elevation
- **Accuracy** — the estimated accuracy of your unit — the circle you are likely in!
- **Location** — the grid reference of your current location.

Setting up like this allows you to use just one screen for the majority of the time you are navigating.

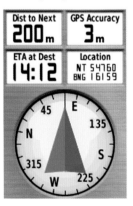  

From left to right: Large Bearing (my preference); Small Bearing; Course Deviation Pointer — if you drift off course this arrow indicates by how much, but is not calibrated for the small changes used in micronavigation and consequently I never use it.

Basic receivers will only have a compass which uses satellites (called a differential compass) and you need to be moving to get a reading. You will often find that the bearing pointer changes direction within a few metres of moving off because of this.

Mapping receivers have a differential compass, but in addition have a fluxgate compass which doesn't require you to move for it to work.

Calibrating the compass

The compass needs to be calibrated and this is easily done by following onscreen instructions and rotating the receiver. You will need to calibrate the compass if:

- you have installed new batteries
- you have travelled more than 160 km since last use
- if the ambient temperature has changed by more than 15° C
- if you are entering an area where you require a high certainty of heading.

When you are calibrating make sure you are clear of anything that could influence a magnetic field, both in your environment (such as wire fences, railway lines, overhead high-voltage cables, underground pipelines) and about your person (such as mobile phones, communication radio handsets, steel wristwatches, pocket knives, belt buckles) – and don't forget any nearby vehicles.

↘ EXPERT TIPS

→ When navigating to a waypoint using the compass page, I shuttle between this and the map page just to confirm there are no obstacles or hazards in between my current location and the waypoint.

→ On older, satnav receivers you must switch the compass off when not required – it consumes a lot of battery power. Modern mapping receivers have auto-switching which does this for you automatically.

→ With more basic two-axis compasses you must hold the GNSS level to get an accurate reading. If you have a three-axis, compass orientation does not matter.

→ Also with older units you need to walk a few steps to confirm your reading: for best results at 4 kph for about 5 seconds.

USING THE ALTIMETER

Altimeters give you your elevation – this is the height you are above the surface of the sea.

This stated height is an important environmental pointer to determine your location when using maps with contours.

Purpose-built barometric altimeters were the mainstay for serious mountaineers, off-piste skiers and wilderness travellers for many years. These determine elevation based on the reduction in air pressure with altitude. However, air pressure does not only vary with height, there are many other factors such as warm air being less dense than cold air and humid air is less dense than dry air. So throughout a journey, barometric altimeters need constant recalibrating to be accurate.

All satnavs will calculate elevation and the accuracy of this reading depends upon the quality of the satellite signals it is receiving and their configuration.

Satnavs generally measure horizontal accuracy to a level of accuracy twice that of altitude when using satellite information alone. Therefore, if your receiver is showing an accuracy of 5 m, then altitude accuracy will be 10 m.

Some satnavs combine both a satellite altimeter and a barometric altimeter. The unit auto-calibrates between these two readings every second – significantly improving accuracy to 1.5 times less that of the horizontal accuracy. So a stated accuracy of 3 m horizontally would indicate a vertically accuracy of 4.5 m.

Confirming location using altimeter height

If the satellite geometry is poor or you are experiencing low visibility, you may wish to confirm the your satnav-stated location.

1. Note the stated GNSS accuracy. Multiply this number by 2 to get altimeter accuracy (or 1.5 if your satnav combines both a barometric and a GNSS altimeter).

2. Read the height given by the altimeter.

3. Add the altimeter accuracy number to this height and make a note of it. Subtract the altimeter accuracy number from this height and make a note of it. Your height is somewhere between these two figures.

4. On your printed map search for the contours between these two readings, you are somewhere along this band.

Free-falling from 4,500 m (15,000 ft), Euan George relies on his combined barometric and GPS altimeter to open his parachute!

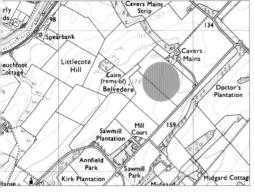

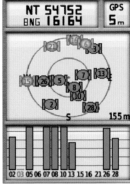

Example

- Stated accuracy 5 m
- 1.5 x 5 m = 7.5 m altimeter accuracy
- Altimeter reading 155 m
- Altimeter accuracy range = 147.5 m to 162.5 m
- Search the printed map contours between 150 m and 170 m

The grid reference put me to the east of the cairn, the red circle created between contours 150 m and 170 m was the area I was in.

↘ EXPERT TIPS

→ If your receiver has a barometric altimeter, at the beginning of every journey calibrate the altimeter if you know your correct elevation. Ideally you should re-calibrate on average every 20 mins, or every 500 m of descent/ascent, whichever occurs first.

→ If you are using a stand-alone barometric altimeter it is essential to frequently calibrate it every time you are at a known height, particularly in changing weather conditions.

When calibrating your altimeter at a trig point, make sure to hold it at the actual benchmark height, not the top of the trig point.

→ Auto-calibration on most satnavs can also manually be disabled if you wish only to use one of the readings. For example inside a tent the barometric reading could be inaccurate or conversely in a canyon where GNSS signal is poor the GNSS altimeter could be inaccurate.

Backup Height

I have coined this term for a technique I use in areas where I want to very accurately mark a feature – such as a junction on a mountain path which could be used as an escape route. If I need to follow the route in an emergency, when I reach the junction, in addition to using other navigational clues (including a waypoint), I confirm that it is the correct junction with a noted altitude.

Altimeters are also an excellent tool to use for:

- augmenting **Slope Aspect** (pp. 154–7).
- helping you maintain the desired height when **Contouring** – and therefore your course.
- a barometric altimeter will show a pressure trend – this is a useful weather indicator if you maintain a constant elevation. Generally falling air pressure means that clouds and precipitation are likely and rising air pressure signals that settled weather is likely.

In addition to providing your current elevation, GNSS can give you:

- minimum/maximum elevation
- total ascent and descent
- ascent/descent rate
- average and maximum ascent/descent rate.

Some units, such as the Garmin GPS60s series, combine both a barometric altimeter and a GNSS altimeter. The unit auto-calibrates between these two readings every second – significantly improving accuracy.

↘ EXPERT TIPS: SATNAV SUMMARY

→ As soon as you have reached a waypoint you **_MUST_** stop the satnav navigating to it, or it will keep trying to take you back there.

→ Whenever you either give or receive a grid reference verbally, repeat it back to the person to make sure it has been correctly understood.

→ Write down grid references, otherwise it is easy to get them wrong!

→ As soon as you press the **Mark** button the location is locked, even if you keep moving – you can change its name, symbol etc., while continuing to travel. This is a useful feature if you are skiing and one of the party falls when you are not able to immediately stop. When you do stop, save the waypoint and navigate back to your colleague. This is similar to the 'Man Overboard' feature.

→ Selecting which north to use: usually found on the Compass screen by selecting menu (also in System/Setup) you find the North Reference— this sets the north reference of the compass as follows:
   True: this is the way lines of longitude run
   Grid: this is the way the grid lines on your map run top to bottom
   Magnetic: automatically sets the magnetic declination for your location
   User: enables you to manually set the magnetic declination value.

USING YOUR SATNAV

What the manufacturers don't tell you.

Because on so many occasions my life has depended on this technology I have asked all of the specialists who I have met how to get the best out of my unit and this is their combined expert advice.

Before you set off

- Make sure you have read **How to Hold a Satnav** (pp. 284–5).
- If you have moved regions verify you are using the correct map, map datum and coordinate/grid reference system.
- Mount on a shoulder strap according to aerial type (see p. 284).
- Switch on your receiver and leave it somewhere with a clear all-round view of the sky (the roof of your car is a good spot), ideally for 15 mins, to get a good fix and collect current almanac data for the entire constellation – which will give you better accuracy for 4 to 6 hours. With predictive ephemeris (Hotfix) the information gathered can be useful for up to several days, which greatly reduces acquisition times.
- Check the stated battery level at the outside temperature – warm cars and jackets can lead to a false reading suggesting more power remaining than there actually is!
- Conserve power – if you don't need the backlight turn it off, the sound too.
- Clear track log and journey statistics if your unit has not already auto-cleared them.
- Calibrate the barometric altimeter (if fitted).
- Depending on circumstances, calibrate the compass (see p. 299).
- Create a waypoint to mark your start.

On your journey

- Give your waypoints simple names so you can both quickly and easily identify them. You can also sort them by the date they were created or edited.
- Keep your antenna dry – a film of water can interfere with satellite reception.
- If navigating in difficult terrain, such as a narrow mountain ridge, set your track recording interval to fine.
- For critical waypoints use the averaging function on your receiver (see p. 316).
- Shield the unit if you suspect multipathing.
- Set proximity alarms before you depart and at a minimum of 10 m if you are going to use them for critical areas.
- Satellites with a D in the bar are SBAS-enabled and therefore provide better accuracy – you need to have enabled your WAAS/EGNOS for this to work.
- If operating in areas of dense foliage such as the jungle, even dense pine forests, fit an external amplified aerial if your receiver accepts one.

At the end of your journey

- Save and name your track if your receiver does not do this automatically.

- Disable the satnav to stop it recording and creating a spike back to your home or where you next turn the unit on.
- MR/SAR members: after an incident, if waypoints were created that may later be needed as evidence give your SD card to the incident commander.
- If you intend to use your receiver again in the next few days leave the batteries in it, as keeping the real-time clock running is helpful to improve acquisition times when predictive ephemeris can be utilised.
- If you will not be using your unit for more than a week, recharge or replace the batteries and do not put them back into the receiver until you next navigate.
- Upload tracks and waypoints to sharing sites for others to benefit.

↘ EXPERT TIPS

→ Print this out and carry it with you at all times. A version of it in PDF is available at my website.

→ Don't forget your lanyard, lithium batteries or backup navigational tools.

CUSTOM MAPS

You can now load any map whatsoever – paper or electronic – onto your satnav and actually navigate on the map.

Working in SAR, this was *the* application I was waiting for. For some time I had been using the sophisticated high-end (read very expensive) Geographic Information System (GIS) to download custom maps onto handheld satnavs involving complex georeferencing, map scaling and difficult file conversions. Now you can do the same for free using Google Earth – and it is so easy to do!

Putting a custom map onto your satnav using Google Earth (GE)

1 Save any map you like as a JPEG file up to 3MB in size. It might be downloaded from a website, such as a university campus map, or scanned from a paper map and saved to your computer as a JPEG.

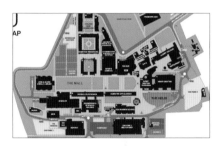

2 Open GE – find the location which your custom map covers and zoom in until the visible area matches that covered by your custom map (JPEG).

3 From the toolbar open **Add/ Image Overlay**. Create the link by browsing to the JPEG file you saved to your computer, select it and give it a name. Keep this window open and move to the bottom of your screen.

4 Beneath the link to your computer there is a slider bar to adjust the transparency of the overlay image – move the slider so you can see through the image you have uploaded on to GE.

5 Deselect all of the visible icons in the sidebar, such as Street View,

Photos, Community Files, etc., by unchecking the boxes found under the **Layers** window.

6 Move the green handles to adjust the corners, edges, centre and rotation of your custom map to fit the GE aerial photograph. Adjust the transparency slider so the overlay image (your custom map) is opaque, and then Save.

7 In the sidebar under **My Places** you will see your file. Move your mouse cursor over it, right-click (ctrl-click on Apple Mac) and 'Save Place As' – browse to where you want to save and call the file whatever you want (saved as a .KMZ format file).

8 Connect your satnav to the computer. Browse the receiver's files (usually this is automatic as AutoPlay starts). Find the folder called 'Custom Maps' and drag and drop the KMZ file you just saved from your computer.

Hey Presto, you can now navigate using this custom map on your satnav.

↘ EXPERT TIPS

→ Any image on your screen can be captured as a JPEG by using the Windows Snipping Tool found in accessories, or cmd-shift-3. on an Apple Mac.

→ If the map has previously been saved in another file format, convert it to a JPEG using graphics editing software such as Adobe Photoshop or GIMP.

→ As you become more proficient use the Location tab on the New Image Overlay window in GE to adjust the draw order. This is used to determine the order in which maps are drawn on your device. If the draw order for your custom map is set to 50 or higher, the custom map will draw on top of the Garmin map.

→ Check the maximum image size you can use with your satnav, as larger images will be automatically downsampled. You can work around this limitation using software such as G-Raster, mapc2mapc or OKMap.

→ There may be a limit to the maximum number of JPGs you can load onto your satnav – check what this is.

PRE-PAID MAPS

Only available on the more expensive mapping receivers, this function creates a truly stand-alone navigational tool.

In most of my trips I only use my mapping satnav to navigate with. However, I always carry a backup basic satnav, and a map and compass.

The two screens that I use the most on my satnav are those that displays the map and the compass.

Topographic maps

All mapping receivers come with a base map and these vary a great deal in detail. Generally to obtain high-quality small-scale topographical mapping you need to purchase it

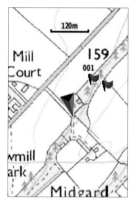

1:25 000 Ordnance Survey map displayed on my satnav (above left). The blue triangle shows my current location and the blue flags waypoints I have created. Compass displayed on my satnav, showing I have 263 m to my next waypoint, and that it will take me 10 minutes 40 seconds, at my current speed, to get there.

separately and this is either available on data cards, such as micro SD cards, which you insert into the device, or transferred by computer to its internal memory.

Mapping such as the UK's OS Explorer (1:25 000) and OS Landranger (1:50 000) and in the USA USGS Topo Maps (1:24 000) give the detail required to navigate on land proficiently.

On the map page you can:

See where you are on the map and equally important where you have been and are going.

Determine your elevation at any given point.

Zoom in or out to view the area around you

See your grid reference.

See your receivers accuracy.

Scroll away from your current location to view other areas of the map.

Display the distance to another feature.

Overlay additional information you have collected or received such as waypoints or routes and view where you are in relation to these.

Choose to navigate directly to either features on the map or the additional points.

Create waypoints and routes on the map.

Loading the maps is very easy: either *plug and play* where you just insert the data card, or the growing downloadable content which is delivered over the web.

Maps using satellite and aerial photography

Google Earth has made satellite and aerial images free to everyone and most people who have access to the internet will have viewed their neighbourhood using it and discovered just how exciting and interesting they are. There are also readily available high-resolution sub-metre colour satellite images from other firms which capture the world in brilliant clarity and detail – it is important to remember that images of towns and cities are usually more up to date than those of more remote areas.

These images can also be utilised on some receivers and displayed on the map screen showing your current location, track, route or waypoints overlaid on them.

- They are an excellent additional facility to help you understand exactly the terrain and environment you are navigating in.
- Often they show paths and trails not on your topographic map, and clearings where you may wish to camp, rendezvous or parking areas, or tree cover for shelter.
- In built-up areas you can view sidewalks, alleys and cuttings not commonly detailed on topographic maps, even individual trees or a small monument.

Advanced receivers also allow you to layer the vector maps on to the topographic maps combining a real-life view of roads, buildings and terrain.

They can either be bought, usually from the receiver manufacturer or downloaded from the internet. The quality is measured in pixels and the higher the number of pixels the greater the detail of the map. Ideally they will have multiple layers of imagery where the image fills in as you zoom in.

↘ EXPERT TIPS

Map orientation

→ This is one feature students find confusing when beginning to use their receiver's map screen. In exactly the same way your orient a conventional printed map to match the features in the environment around you and your direction of travel, your device can do this too – this is called **Track Up**.

→ Additionally it can orient the map as if you are looking at a printed map on a desk where all the text is the right way up and level – this is called **North Up**.

→ The confusion arises as often, especially if you stand still, when the map is set to Track Up it will jump around on the screen. It does this because of the inaccuracy of your receiver, constantly moving your location a few metres in your circle of accuracy, added to which the GNSS compass alone will not work when standing still. So when standing still or planning with the onscreen map, change to **North Up** and when moving again, **Track Up**.

BACKUPS

You will, over time, gather a considerable amount of data from your waypoints, tracks and routes and this will be valuable to you.

Equipment fails, we lose items, we make mistakes – like pressing the **Delete** button, when we meant to press **Save**! So backup your hard-earned data frequently. I have disciplined myself to do this after every trip. There are various ways you can backup your data:

Direct to your computer

The most common method is to transfer files from your satnav to your computer. Connecting via a USB, or wirelessly, you can copy directly into a folder of your designation – screenshots, icons and GPX files. Usually this can be achieved by simply using '**copy and paste**'.

Some manufacturers also supply software which enables you view your data on their maps and if you export the MAP files, which some receivers use, you will have to install the manufacturer's (usually free of charge) synchronisation software onto your computer – this will give you the option to export your route data as GPX files, MAP files or Tab delimited text files.

- GPX files: choose this option if want to view your routes or points of interest (POI) in other mapping tools such as the Online Route Planner and Google Earth.
- MAP files: choose this option if you simply wish to back up your files since MAP can be copied straight back onto your satnav more quickly than other route and POI files, and the rich text formatting in waypoint/POI descriptions is also perfectly preserved.
- Tab delimited text files: choose this option if you want to manage your routes or POIs in spreadsheet programs such as Microsoft Excel.

To your digital mapping

You can export this information to your digital mapping software. This is my preferred method and I only back up onto another media, usually a CD or USB dive, when I am travelling and do not have access to my digital mapping program.

Open your digital mapping program and select the GNSS/GPS tab to initially configure your receiver. This sets up the protocol so that your receiver can 'talk' to your software. Most software recognises most receivers and this process is usually straightforward; you simply select which model you own from the menu and the program does the rest for you.

You can choose to install your waypoints, tracks and routes separately, or all of the data at once. You can then choose to either display this information on the maps of your digital mapping or hide it.

On the internet

There are numerous websites where you can share your data, and therefore effectively save it, from sites which share walks and trails, to **Geocaching** and manufacturers' use websites. You can do this either direct from a file you have created on your computer to store the information or using software provided by the website.

Receiver to receiver

A perfect option for when you are out in the field is to send your data to a colleague's/friend's receiver wirelessly and then retrieve this information when back at base/home.

Shareware

A lot of software has been created by interested users of this technology that can easily be downloaded via the internet. Shareware is either free or the programmers ask you for a small voluntary contribution.

An example of such programs is GPS Utility (PC only), which is an easy-to-use application that provides management and manipulation of GNSS information. The program converts between different map datum and many coordinate formats (lat/long, UTM/UPS, country grids etc.). You can use it to:

- Transfer data to/from a satnav.
- Store the data in PC files in one of several text formats.
- Convert map datum – when transferring tracks/waypoints recorded in say OSGB36 Airy Spheroid (UK National Grid) onto Google Earth, which uses WGS84, there will be a discrepancy between your recorded location and the one displayed: GPSU corrects this.
- Information can be filtered in various ways and waypoints sorted according to specified criteria. Route and track statistics are available and can be transferred into other programs for analysis (spreadsheet programs).
- Using a scanned or digital bitmap you can digitize waypoints, routes and tracks.
- You can plot your GPS information as a map and add map annotations in text or image form.

↘ EXPERT TIPS

→ With a small minority of receivers, especially those which use MAP format files, you will need to export data using the manufacturer's synchronisation software onto your computer first as GPX files and then import them.

GEOCACHING

An outdoor game that helps participants to learn and refine their GNSS use in a fun way.

Geocaching is a high-tech version of Hide and Seek. The premise of the game is very simple: players hide containers, called geocaches, anywhere in the world for other players seek them out. These geocaches are usually small waterproof plastic boxes that contain a logbook which is signed and dated by the finder, who then returns the box to its hidden location. Often they contain novelty items of little value such as trinkets and toys, the idea being you take one and leave another. It is a rapidly growing sport and there are millions of caches across all continents ... even including Antarctica.

A geocache is first hidden, not buried, and then using a handheld satnav a waypoint is created for its location. Anyone can do this and it is not a requirement of the game that every player does but obviously it increases the enjoyment of the sport the more who do.

Players download latitude and longitude coordinates from a website for the area they live in and then put them into their handheld satnav. The cache webpage may include an encrypted hint and previous finders may have uploaded photos or their own clues.

Getting started

Type 'Geocaching' into an internet search engine – this will point you to one of the many sites dedicated to this sport.

❶ Enter your location, usually a postcode is enough. Select a geocache to find – some are an easy adventure in a park, others are located on mountain tops – and note its coordinates.

❷ Enter these coordinates into your satnav using **Send to GPS** function as it reduces the chance for data entry error. This is called forward projecting a waypoint and used frequently in navigating using GNSS.

❸ Find a topographic map of the area, begin with the online maps to get an idea of the area and then decide to supplement with a detailed paper map and the mapping on the GNSS if it is installed. Relate your waypoint on a topographic map, then study the map to gain an appreciation of the level of difficulty you will encounter in the terrain you have to cross.

❹ If it is quite a long way or through difficult terrain you will have to create a route. Before you set off on your journey inform somebody where you are going and what time you expect to be back.

❺ Create a waypoint at your start point, possibly your car, to ensure your safe return.

↘ EXPERT TIPS

→ If you need to enter your home location in lat/long you can either check them on your satnav or obtain them on Google Earth (see **Digital Mapping**).

→ If the geocache waypoint is available as a GPX file on the website you can download it straight to your satnav and digital mapping if you have it.

6 Navigate relating the information on your satnav about distance and direction to the geocache with your map (either paper or installed). Once you are close to the cache location follow the GNSS bearing indicator (arrow).

Geocaching: GNSS technology soon becomes familiar in children's hands.

↘ EXPERT TIPS

→ If you use a website to convert your latitude and longitude coordinates to a local grid reference they may be a discrepancy in the exact position because the local grid may not be based upon the WGS84 datum.

→ Most geocaches are in plastic boxes although sometimes they can be disguised as rocks. Any container can be used so long as it is weatherproof and not too large.

→ As the accuracy of GNSS has improved and players become more skilled a variation of the game called micro-caches has started. These are containers often no larger than a playing card and consequently more difficult to find.

→ Geocaching is an excellent way of introducing the technology to children.

This game will help develop most techniques that you need to perfect in order to become a proficient navigator using GNSS, from simple tasks such as marking a waypoint to more involved ones like changing the position format on your satnav to lat/long and then relating this to a regional map.

Geocaching also highlights the main limitation of this technology: estimating distance and obstacles that you may encounter. All satnavs display distance as the crow flies and take no account of any obstacles you may encounter en route.

The distance does not take into account the elevation of the ground, so if there is ascent/descent involved the distance to travel will actually be longer than that stated by your satnav. The situation is, however, improving, as it seems likely that elevation data will be incorporated into many satnavs in the future

Paperless geocaching

Many handheld satnav units now have special pages (dispays) dedicated to Geocaching that can store up to 5,000 locations, cache descriptions, ratings, and recent log info can be stored eliminating the need for you to take paper notes with you.

ADVANCED USE

CREATING ACCURATE WAYPOINTS

In most situations the accuracy of waypoints you mark with your receiver is acceptable for everyday navigation.

However, sometimes you may need your waypoints to be as accurate as possible: for example, at a critical turn-off on a mountain path or an equipment stash. There are several techniques you can employ to improve accuracy – averaging is the most accurate.

Averaging a waypoint

If your receiver has the **Averaging a Waypoint** feature you should use this. The receiver starts multiple recordings of the location every second and averages these. The optimum time required to do this is 7–10 mins. Leave the unit stationary, in the correct

Taking successive readings of waypoint: mark your position with a stone so you return to exactly the same spot each time.

→ You can also return to the same spot on another occasion to collect more data to improve the average.

→ If when recording the position manually one reading is significantly different from the others, ignore it.

antenna position, such as on top of a rock, with a clear view of the sky. You can return to the same spot and open the waypoint you originally created and average it again; the more times you do this the greater the accuracy. For optimal results, at least four to eight samples should be collected at least 90 mins apart – allowing the satellite constellation sufficient time to change – or ideally over different days.

I have achieved sub-metre (300 mm) accuracy in Scotland when trialling this technique at very precisely surveyed landmarks. So in worst case scenario I would only be 1.5 m away when returning to find this waypoint.

Manually averaging
You can also manually average a waypoint. Create the waypoint as usual and record its grid reference. Fifteen minutes later repeat this process. Do this six times, 90 mins in total. Add the six **easting** readings up and divide by six to get the average **easting** and do the same for the **northing**. In the area that first read NT54817 162225 the recordings were

I	NT 54817	16225
ii.	NT 54819	16233
iii.	NT 54820	16230
iv.	NT 54823	16229
v.	NT 54823	16227
vi.	NT 54822	16232
Totals	NT 328924	97376
Divided by 6	**NT 54820.6**	**16229.3**

In fact, the precisely surveyed location is NT54820.3 16230.5

- The first reading was out 3.3 m easting and 4.3 m northing: consequently, on the basis of this one reading, the maximum discrepancy between my actual position and my assumed location when searcing back to a waypoint would be 21.5 m.
- The averaged reading was out 0.3 m easting and 1.2 m northing: consequently, using the averaged figure, the maximum discrepancy between my actual position and my assumed location when searcing back to a waypoint would be 6 m.

PROXIMITY ALARMS

A proximity alarm is a waypoint which has an invisible radius around it – if you enter into this space an alarm sounds on your receiver.

They are excellent to use for marking spot hazards, such as potholes, or creating a virtual wall of protection against linear hazards, such as a cliff edge.

Creating proximity alarms: a cliff edge (left) and potholes in a forest (right).

Creating a proximity alarm

There are four ways to create proximity alarms:

- when you are at the location
- forward projecting a waypoint
- on your digital mapping
- receiving an alarm somebody else has created either via the internet or sharing wirelessly unit-to-unit.

At the location

1 Create a waypoint in the normal way: if this is a critical waypoint create an accurate waypoint (See **Creating Accurate Waypoints**).

2 When you have created the waypoint from the Main Menu select **Proximity Alarms**.

3 Select **Create Alarm**. You will now be given a list of options: Use map/Recent finds/ Waypoints/Photos … and other points of interest.

4 Select **Waypoints**. Choose the waypoint you have just created. Select **Use** and you will be prompted to enter the radius of the alarm. Enter the radius and select **Done**.

5 You can now **Quit** or, to review the alarm, highlight and select it to change its radius, view on the map or delete it.

Forward projecting a proximity alarm

1 From the Main Menu select **Proximity Alarms**. Select **Create Alarm**. Select **Use Map**.

2 Move the cursor on the map to where you want the alarm, press **Enter** and you will be prompted to enter the radius of the alarm.

3 Enter the radius and select **Done**.

4 You can now **Quit** or, to review the alarm, highlight and select it to change its radius, view on the map or delete it.

Using simple proximity alarms

The most obvious use of these handy little features is to mark an object or area you wish to avoid. It could be a small pothole, not marked on your map, a boulder, again not marked on your map but which could catch you out at night, or an area of land which has become flooded or is difficult underfoot. They have many clear-cut uses.

I usually add 15 m to the radius I would like to protect. So if you wish to be warned of an upcoming pothole 10 m before you reach it make the radius of your alarm 25 m. If you are marking an area of marshy ground that is 35 m in diameter create a proximity radius of 50 m. This gives you both time to react and allows for any GNSS inaccuracy (see **Accuracy**, pp. 286–7).

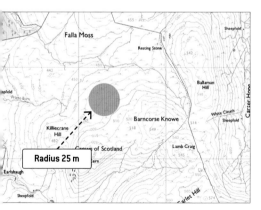

Pothole marked with a 25 m proximity alarm.

On SAR missions they are an excellent tool for:

marking the **Locus** – sharing this with other responders so that they do not enter and contaminate a potential crime scene.

marking dangerous wreckage, such as jagged metal parts from an air crash, sharing this with other responders so that they do not injure themselves.

marking items found during the mission so other responders do not disturb the area further.

Creating Virtual Walls with Proximity Alarms

I developed this technique for use by MR teams and now instruct it to all professional groups, from law enforcement agencies to military personnel.

By overlapping proximity alarms you can create a virtual barrier of any shape and length, which, if approached, will cause your satnav to sound an alarm.

For instance, on a narrow mountain ridge you could create a wall of proximity alarms either side of the ridge, so that in poor weather an additional safety measure to good navigational procedure would be the walls of alarm either side. The same applies for linear areas of danger, such as a fast-flowing stream created after flash flooding, or a cliff edge.

Set proximity alarms up before you depart. Space them at a minimum of 30 m intervals – 50 m is the ideal.

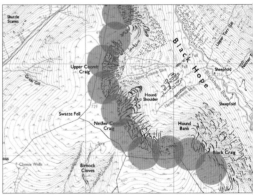

Such alarms could mark an entire group of buildings where a perimeter cordon has been established at a scene of crime, or the perimeter of a minefield or line of unexploded munitions dropped from an aircraft on a battlefield.

They can either be created *in situ* by physically creating a waypoint at the spot and converting it to a proximity alarm, or they can be created using digital mapping and downloaded onto the GNSS handsets of the individuals who will be travelling to the location. The number of applications for this technique is limitless.

GNSS IN EMERGENCY MANAGEMENT

This is an overview of the training I deliver to MR and SR teams which is a two-week long course.

atnav is an invaluable tool when used in emergency management, such as the work conducted during the 2010 Pakistan floods, to major terrorist attacks. I have written his section as an introduction for teams/agencies involved in the management of mergency situations. It is by no means an exhaustive list and you will add your own nput as your systems grow.

The combining of GNSS data with GIS for use in emergency management is nothing hort of a revolution, providing for ground-breaking new strategies and tactics. Gone re the days when GNSS/GIS may have been useful, optional tools to consider in mergency planning and response. The technologies have now become so reliable in nabling us to plan, conduct and expedite missions more quickly and safely that we ould not be fulfilling our duty of care to our response teams, the casualties and the eneral public at large if we did not employ them.

Bespoke emergency management digital mapping is now readily available. Some reated by the emergency service themselves, such as the missing person mapping rogram, Mountain Map, developed by Scottish Mountain Rescue, or EmerGeo Mapping hich can be integrated with Google Earth.

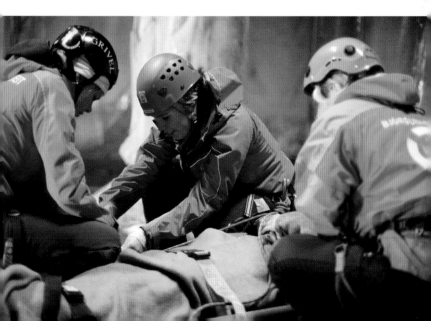

These all work on the foundation of a conventional digital mapping program with additional features and benefits specific to the emergency sector. At the same time, conventional digital mapping should not be dismissed as a tool, as it is now of such high quality and flexibility that it can be tailored to the emergency services' specific needs.

There are five stages/levels of implementation.

Stage I: System Selection and Management

Talk to similar groups about their experiences and get feedback – see what bespoke software may be available for your given type of emergency management.

Contact other emergency services who you work with in your area and ask what data they have compiled and if they will mutually share it with you to save duplication of effort. Ideally, if their software meets your requirements you should consider buying the same, although information is readily interchangeable between different GIS.

Select a digital mapping program which has high-quality topographic maps and satellite images/aerial photography available for your area of responsibility and undertake to assess the software before going live:

- determine the quality and support for the product
- does the system have sustainability and longevity?
- what are the limitations and constraints?
- assess vulnerabilities, including the security of data held
- outline costs and benefits.

Choose receivers that are compatible with this software or if you already have receivers, *vice versa*.

Key features

Real-time asset tracking is an invaluable benefit, from vehicles to people. Receivers are being built which have the capability to live-link with control GIS and this should be a serious consideration.

Can the system accept a map produced in the field? With modern communications the map can be sent electronically, georeferenced manually or automatically for instant use at an incident. See **Custom Mapping** (pp. 305–6).

Nominate a person or small group who will have overall responsibility for the management and running of the software: deciding what type of information will be uploaded/downloaded, how it is collected and how it is both subsequently used and distributed. Teams that have not done this usually find that the system can soon become unwieldy to manage with superfluous information and duplication of effort consigning it to the NTB (novel tool bin – all teams have these!).

Train every responder and member who uses a receiver to a set standard: have clearly defined standard operating procedures and work methods. The system is only as reliable as the inputted data.

Stage II: Base Operational Map

Define the geographical boundaries of the territory you are responsible for and add standard operational information, locations such as:

- headquarters – include telephone numbers, details of where keys are held and alarm codes
- operating bases
- car parking facilities
- vehicle storage
- refuelling points
- equipment stores.

Enter the same information (where available) as above for emergency services you cooperate with.

Define routes for teams outside your territory who may be called to assist you to the above location.

Mark areas by landowner, with their contact details for access, permission to train, and so on.

Stage III: Building the map

This requires the commitment and involvement of the whole team who are going to actively collect raw data in the field. This can be combined with training sessions and also completed during the team member's leisure time. This is especially applicable in MR as most team members are keen mountaineers and walkers so will do this for pleasure. The data collected via this method varies considerably from team to team, service to service.

Data processing must include:

- Standards and specification
- Network configuration
- Data acquisition
- Data analysis

Building the map.

- Data backup
- System failsafe.

Basic data
This information is simply obtained by either creating a waypoint or recording a track.

Gates
In my own MRT we cover an area with a great many fences and walls so vehicular access depends upon access through gates. Gates are not marked on OS Landranger or Explorer maps so we waypoint them whenever we come across them, whether they are new or old, and mark them as locked, open and detail what can pass through them – a vehicle or pedestrian.

Small bodies of water such as pools and ponds
On OS Landranger maps (1:50 000), bodies of water less than 50 m in diameter, and on OS Explorer maps (1:25 000) less than 8.25 m in diameter, are not depicted – yet these can significantly change the parameters of an area to be searched and how that search will be conducted.

Other potential hazards
- Small areas of land subsidence
- Industrial areas
- Chemical stores
- Munitions/inflammables stores
- Pipelines
- Coastal pollution
- Specific radiation hazards.

Kit/equipment stashes, including ropes, pulleys etc.

Graded data
This information requires the person collecting it to access the route/waypoint and measure it against a predefined set of criteria.

Tracks and paths and accessibility:
- have they changed course, are they new or disused?
- are they passable for: a team member to walk through?
- to walk with an ambulatory casualty?
- to carry an empty stretcher up?
- to carry a stretcher with a casualty down?
- a 4x4 vehicle?

This information is layered so that during an incident, if you need to identify all 4x4 vehicle access this can be exclusively selected.

Important: Potentially, hundreds of waypoints will be recorded by many different people, therefore it is essential to differentiate between who collected the information and exactly what they marked. To do this, agree a common format for the different types of waypoints in your area, in Scottish mountain rescue these are:

Type of Waypoint	Abbreviation	Type of Waypoint	Abbreviation
Cellular Mast	CM	Electricity Pylon	PY
Fence	FC	Road	RD
Gate	GT	Scree	SC
Kit stash	KT	Stream	ST
Path	PT	Track	TK
Pothole	PH	Wall	WL

Information about the waypoint can be added in the waypoints notes, such as an access code for a door in a building, the combination lock on a gate-lock or a telephone number. Pictures can also be attached to the waypoint to help identify exactly what the responder should expect to see.

Most satnavs allow the input of special icons to designate waypoints, the handset will then automatically default to this icon. Custom icons can be created to identify each individual and the team: for example, for one team it could be blue circles with a number that relates to each team member. Liaise with other agencies you work with to agree colours for each team.

Pool this bespoke data collection with other agencies in your area that use digital mapping.

Stage IV: Incident Management – data collection, synthesis and analysis

Marking the incident information including and not limited to:

- Descriptions, location, date, time of incident, time of call.
- PLS – place last seen

- Last known point
- Incident command post
- Rendezvous point
- Parking
- Defining areas to be searched
- Layering statistical zones around the incident.

When briefing teams, upload the boundaries of their search area to their receiver, along with waypoints or tracks that will help them in their assignment.

- Deployment of team members, vehicles, equipment – where and when.
- Create 'Transportation' maps. These could be either local maps for getting to and from assignments or larger-scale versions which could be emailed to incoming teams to get them to the search.

At the end of their search/mission, team groups should upload collected information to designated group leader who, upon return to HQ, immediately downloads from receiver the data they have collated of recorded tracks and waypoints (name each individually) and debriefs them while looking at the downloaded data on the computer. Assign the track to the correct type such as: Dog air scent/Hasty team/Line search.

Collate and name all waypoints of items marked such as:

- clue found
- georeferenced photographs
- georeferenced movies
- dog alert
- danger area
- the locus.

Can any of the above information be transmitted live to provide real-time tracking and navigation (3G/GPRS/GSM/SMS/Satellite/Tetra/APRS) and if so determine which is relevant to control the input of data flow.

Using Copy/Paste, add these tracks and waypoints into any Situations maps, Master maps, and Clue Log maps as necessary, and save these changes.

Study tracks created while searching for potential blind spots or incidents where the searcher had started to move too quickly for effective searching (greater than 4 kph) – this often happens towards the end of a search.

GNSS evidence at Scenes of Crime (SoC).

The author has initiated a project working with universities, law enforcement agencies and the various government departments that control the space segment of GNSS, to establish 'indicia of reliability' for data collected at SoC to be admissible evidence in a court of law.

Based upon the actuality that any GNSS satellite constellation configuration is continually changing, at present the chances of replicating the exact geometry of satellites visible for a given location, at a given time and date are almost non-existent.

If the incident is a potential scene of crime the **Locus** can be marked with a proximity alarm and this is then shared this with other responders so that they do not enter and contaminate a potential the scene of crime.

Creating proximity alarms for items found near the incident or on the search so other responders to not further disturb the area.

Incident locus

Track of first responder

Proximity alarm

The track of the first responder at the scene of crime will show the path where possible contamination could have occurred for forensics.

This same track can be used by all responders who have to visit the scene to enter and exit on.

Stage V: Post-mission analysis and incident log

The waypoints and tracks collected in the field are contemporaneous evidence of each responder's activity and should be backed up and stored securely.

The power of learning from these records cannot be underestimated: as a system it can provide scenario planning, action plans, 'what if' scenario testing and management of real-time problems:

- determine strategies for mitigation, preparedness, relief and recovery
- identify future training requirements
- develop incident specific management strategies.

The only limitation to this technology's application in Emergency Planning and Management is our thinking!

JAMMING AND SPOOFING

The very weak nature of the GNSS signal means that the system is open to interference from external parties.

Jamming – flooding an area with radio signals on the same frequency and modulation that your satnav uses to receive the satellite broadcasts.

Spoofing – transmitting a signal which purports to come from a satellite when in reality it does not, so the data sent is incorrect.

Unfortunately I must write about the potential for interference with the satellite signals and satnavs where the possibility exists to corrupt or block data, resulting in either no location data being displayed or false readings. The Achilles heel of GNSS is that its signals are extremely weak – in effect, each satellite transmits data at less power than a car headlight, from more than 20,000 km above the earth's surface.

Most Western governments have experimented with disrupting GNSS signals to prevent enemies of the state using them for navigation. The British Government's Air Warfare Centre conducted a major GPS jamming exercise over northern Scotland in November 2009, and The Homeland Security Institute, a research arm of the US Department of Homeland Security, conducted spoofing tests in the summer of 2010. These are a perfectly legitimate and responsible exercise for the defence of the realm; however, if you are unaware of them and are navigating in a test area your satnav will not function. These trials are pretty infrequent – only three were published in the UK in 2009.

A more insidious potential for jamming exists with criminals – the motivation being that high-value assets are tracked with GNSS (such as armoured cars) – confusing the tracking signal could spell a successful heist. In addition, satnav-based pricing for toll roads and road usage charges could be spoofed, and a company's employees may even be tempted to block the tracking devices fitted to many company vehicles these days.

Nature
Very occasionally natural phenomenon has been seen to effect GNSS reliability and accuracy. These include: sunspots, solar flares, solar maximum; multipathing; and severe, low-latitude equatorial storms.

↘ WARNING

The possibility exists that your GNSS signals could be corrupted and given that it is highly unlikely that you will know if this is happening, it reinforces the fact that it is essential you carry backup navigation in the form of a map and compass.

DIGITAL MAPPING AND GNSS

In isolation these are great tools; together the only limit they have is our imagination.

For a great deal of the time I used a handheld satnav to navigate my way around the world and when I got back to base I would switch on my computer, start my GIS and plan tomorrow's adventure. The first time I overlaid a route onto a map was a revelation – I could see what speed I had travelled and where, transfer points of interest I had marked on my satnav direct onto the map and over time the maps became such a real part of my environment I comfortably knew what I could and couldn't achieve on a day's journey and planned all of my trips more accurately and therefore more safely too. If you do no more than this, digital mapping will significantly improve your ability to competently navigate.

The truly exciting part begins when you push the boundaries of this technology, from sharing routes and waypoints via the internet with friends and colleagues, to creating customised maps for everything from leisure, to SAR missions.

For all users, from athletes to SAR team responders, location data can be combined with other data to better understand your personal limits and your environment, consider heart-rate data, altitude, air temperature, etc.

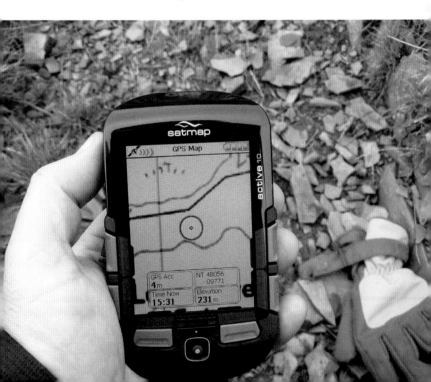

As a new technology develops, it creates specialist offshoots, all of which have similar names or designations and the whole area becomes complicated by similar-sounding terms: 'geographical information system', 'digital cartography', 'mapping application', 'geospatial information system', 'route planning and GPS tool', 'digital map software' and so forth.

Put simply, the software we are going to discuss and use joins together maps and databases of geographical information. It allows you to capture, store, analyse and manage information that is linked to a specific location using digital versions of maps on your computer, laptop, pocket PC, mobile phone and/or satnav. You will use it to plan your trips and to download and upload data to your satnav or other mobile device. In this manual this is referred to as **Digital Mapping**.

Digital mapping should be a serious consideration for all navigators irrespective of whether or not they possess a satnav. First, we are going to look at Google Earth, which is the free-to-use program I recommend you use before buying your own digital mapping.

Google Earth

This digital mapping application has revolutionised the way people explore their world. It was a concept of such scope that during early planning meetings many executives believed it could not be achieved in their lifetime. Google Earth (GE) is continually evolving with an already massive database (many terabits of data) that displays satellite and aerial photography of the earth's surface mapped onto a 3D, WGS84-based model of the planet. It enables users to view any point of the planet, zooming in, flying over – looking perpendicularly down or at an oblique angle, with perspective.

It also combines maps, 3D building photographs and detailed photographs along many streets and tracks using a technology called Streetview where vehicles from

View of San Francisco Bay in Google Earth – showing 3D terrain and added man-made features along with mapped photogaphy.

↘ QUIRKY FACT

→ Believe it or not the world's first true geographical information system (GIS) ran in Ottawa, Canada by the Department of Forestry and Rural Development in 1962!

pecially adapted cars to snowmobiles take 360° photographs along the route. In ddition, users of GE are encouraged to add their own location photographs.

The degree of resolution available depends on the area being viewed and its likely nterest and popularity – but most land (except for some islands) is covered in at east 15 m of resolution (each pixel represents 15 m x 15 m on the ground), while nost Western cities are represented at a higher resolution (right up 0.15 m) – e.g. ambridge, UK, Melbourne, Australia, or Las Vegas, USA.

The program's 'Layers Pane' allows you to determine what sort of information ppears on the map. For instance, you can choose to display roads, hospitals, grocery tores, restaurants, golf courses, bars … even crime statistics!

It operates across on the leading operating systems, including PC, Mac or Linux, and n many smart mobile phones.

To get started, go to **earth.google.com** and download the latest version of the oftware for your intended operating system for free.

The basic features are intuitive: to search for a place, ou simply enter information for the location such as s place name, post/zip code or grid reference and GE /ill begin slowly zooming in on your destination. After moment, you'll see a satellite image of the place you ntered. Controls in the location window let you zoom n or out, move north, south, east, and west and save pecific locations on the map. This movement across he photographs is called flying and you can fly from ne location to another, even get directions.

Type into the search box your post/zip code and view /here you live. Move the cursor over your home and use he zoom button to get a more detailed view. The grid eference, in latitude and longitude, of your home is at he bottom of the screen.

> '*OK, I am going to have to mention map datum again here! Google Earth uses WGS84 and when you transfer coordinates from your satnav you must make sure they are also WGS84 otherwise their exact location will not be accurately displayed in Google Earth.*'

GE is compatible with many satnavs: you can import waypoints, tracks and routes irectly by simply connecting the satnav to your computer and importing the files from he GE Tools menu. Alternatively you can copy the files onto your computer. Read the nstructions that came with your satnav for exporting data from your device to a file on our computer. You can also use third-party software like GPSBabal to download the PX file to your computer.

Once you have loaded your data into GE, you can edit the waypoint placemarks and rack paths, and add more information in the balloons. To edit a feature in GE, right-click n the feature in the 3D viewer or the **Places** panel, and choose **Properties** (on a PC) or et Info (on a Mac).

GE can also be used to search for schools, parks, restaurants, and hotels. You can tilt nd rotate the view to see 3D terrain and buildings, save and share your searches and

GNSS

favourites and even add your own multimedia – photographs can be geotagged (i.e. they appear where they were taken), videos can be linked and annotations added to specific features.

To show the waypoints, tracks and routes from other satnavs save them to your computer and then open them in GE from the File menu – remember to specify the correct file extension (GPS, LOC, MPS files) on the button next to the File name in the window GE opens.

Free regional mapping programs

Search the internet for free regional mapping and see if it meets your needs before deciding to buy proprietary digital mapping.

www.maptogps.com

This site allows you to create a route anywhere in the UK on maps from Ordnance Survey, including 1:25 000 Explorer and 1:10 000 Landplan maps, and generates GPX files which can then be uploaded to your handheld satnav. It also allows you to edit GPX files you have already created.

To get started

1 Enter a location in the text box and click find. If more than one location matches what you have typed in, you will be presented with a list of locations.

2 Click on the correct location from the list and you will go to the map page showing your location.

3 Click on the map to draw a route.

4 When your route is done, click the create GPX tab.

5 Download this file.

Editing routes already created:

1 Click on the upload GPX tab to upload your GPX file.

2 You can now edit the route on the map page.

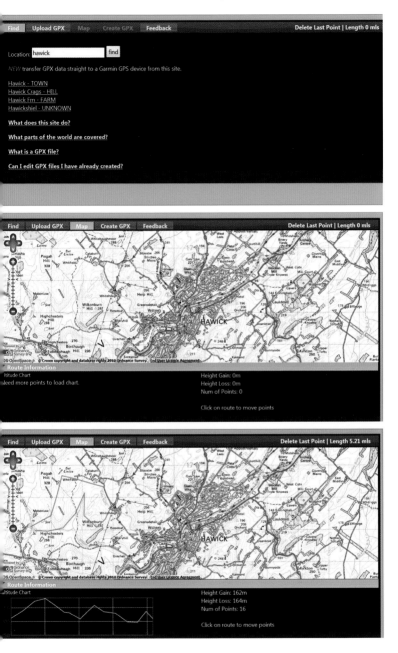

The total distance for the routes is displayed, along with the height gained and lost, plus an altitude chart.

A similar global program is also available at **www.openstreetmap.org**, which is a free editable map of the whole world and is made by users of the website. It allows you to view, edit and use geographical data in a collaborative way from anywhere on earth.

Proprietary digital mapping

There are numerous manufacturers of digital mapping programs designed for the leisure market. I have extensively worked with Anquet, Fugawi, Memory-Map, Tracklogs and Quo. My preferred choice is Anquet.

Anquet is intuitive to use, has unparallel 3D capabilities, great customer support, web-based tutorial videos and quality digital maps from the likes of Ordnance Survey, Philips, Harvey and Getmapping aerial photography in the UK. International mapping is just becoming available at the time of writing, starting with France, Ireland and America – check their website for the latest maps available. You can download the Anquet software for free from their website, including some sample maps, allowing you to get to grips with how the system works – you can directly test it with your satnav, plus any Windows Mobile pocket PC or smartphones you wish to use it with for uploading and downloading data and maps. If you then choose Anquet as your digital mapping you only pay for the digital maps you select and once you have bought these they can be downloaded, including tracks, waypoints and routes, and installed onto portable navigation devices such as smartphones and satnavs.

The main reason to buy bespoke digital mapping is the flexibility it gives you and the high-quality, large-scale maps you can use. You can display maps such as Ordnance Survey 1:50 000 and 1:25 000 or USGS 7.5 minute maps on your screen, and view them in exactly the same as you would a printed map. Digital mapping makes access to these maps quick and effective and allows you to overlay your own information onto them.

You can move across and search the map seamlessly, unlike paper maps where you have to change from sheet to sheet and try and match them up if your route is on two maps. You can also quickly find any location on a digital map either by place name

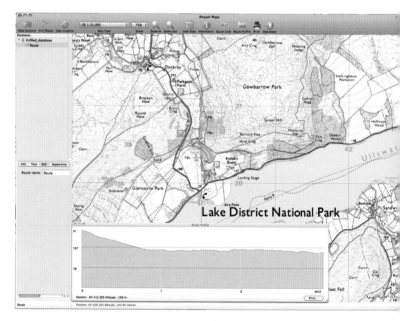

Anquet: screengrab showing a route in the process of being plotted.

ar grid reference. Once at a location it is easy to change to other maps that you have access to, or to view maps simultaneously. This is amazingly powerful functionality, allowing you to move to more detailed or less detailed maps at the click of your mouse. Comparing map types can also be very enlightening. For instance viewing aerial photography alongside detailed topographical maps can often reveal features not visible on the topographic map, yet without the aid of the topographical map, the aerial map is extremely hard to 'read' as it lacks the symbols and conformity of a map produced by a cartographer.

Guide to Anquet

1 Follow the prompts to install the software and register your Anquet user account.

2 Free demonstration maps will install automatically.

Moving around the map

1 Use the cursor keys to scroll, or

2 Hold the left mouse button and the map will move as you move the mouse.

Zooming in and out

1 Use of the + and − icons on the toolbar, or

2 Select **Zoom In** or **Zoom Out** from the map menu, or

3 Select a zoom % from the drop down menu on the toolbar, or

4 Roll the mouse wheel forwards to zoom in and backwards to zoom out.

Changing map type

1 Select the **Change Map** option on the map menu (map types available at your current location are listed in black text in the menu), or

2 Click on the drop down list of map types on the toolbar.

Finding places by name

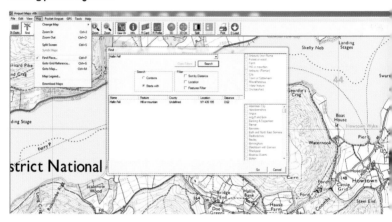

❶ Select **Find Place** in the map menu, or click the **Find** icon on the toolbar.

❷ Enter a name in the top window and click **Search**.

❸ Highlight your selected name in the lower window and click **Go**.

Finding locations by grid reference

❶ Select **Goto Grid Reference** in the map menu or click the **GoTo** icon on the toolbar.

❷ Choose the grid type (UK Nat. Grid, Lat./Long., UTM, etc.).

❸ Enter grid location in box.

❹ Click **Go** to move to the location or **Add Waypoint** to drop a waypoint at the location.

GNSS

336

Drawing a route

1 Right click with the mouse and select **Start Route**.

2 Create the route by moving the cursor and left-clicking at each change of direction.

3 To draw a continuous curving line, hold down the left mouse button and follow the route with the cursor.

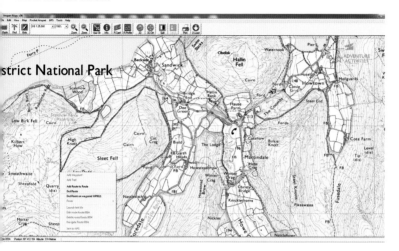

4 Release the button and you revert to the single click and move routine.

5 To end a route, right click and select **End Route**.

Viewing route information

1 To view route information select Information from the **View** menu or click the Info icon on the toolbar.

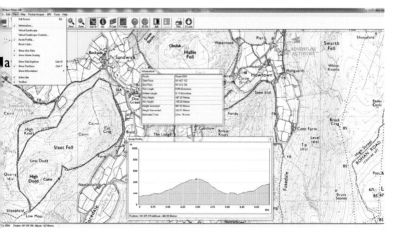

2 To view a route profile select **Route Profile** in the view menu, or click the **R. Profile** icon on the toolbar.

Printing maps at various zooms

1 Select **Print** in the file menu or click on the **Print** icon on the toolbar.

2 If the zoom level in the print box is set to 100% the map will be printed to scale so on a 1:50 000 scale map, 2 cm will represent 1 km.

3 If you print a 1:50 000 scale map at 200%, 4 cm will represent 1 km.

Printing maps to scales

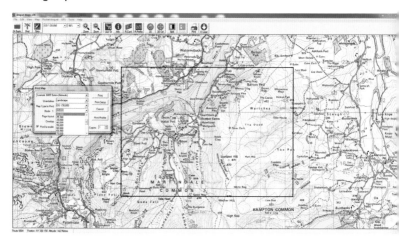

1 Tick the **Print to Scale** tick box (the zoom options will change to scales).

2 Select required scale from the drop-down menu.

3 You can print any map data to any scale (e.g. a 1:50 000 map can be printed at 1:25 000 scale).

Printing maps on multiple sheets

1 A black box outline shows the area that will be printed on an A4 sheet.

2 You can move the map around to line it up in the box, then press **Print**.

3 If you cannot see the black box, the print area is larger than the current screen display.

4 To see the black box, zoom out using the negative magnifier on the toolbar or increase the Zoom level in the print box.

5 If the route to be printed is too large for an A4 page at the desired scale, select a set of **Tiled** pages or **Best Fit** in the Page Layout menu and a series *of A4 pages will be displayed and can be printed*.

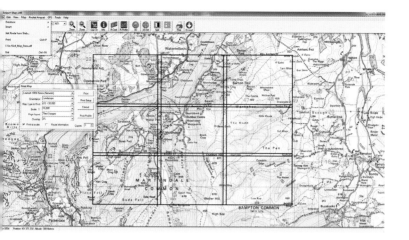

Using split screen and screen synchronisation

You can view any pair of maps for which you have data, and one or both can be in **Virtual Landscape** mode. You can work in each map independently. The active map has a highlighted border.

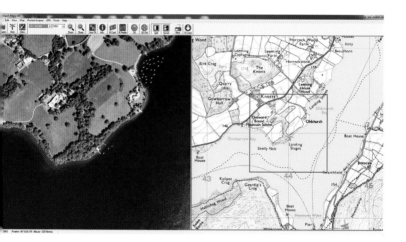

- Select **Split Screen** from the Map menu or click the **Split** icon on the toolbar.

- To synchronise the two maps select **Synch Maps** in the Map menu or click the Synch icon on the toolbar.

Using the Virtual Landscape TM 3D map viewer

- Select **Virtual Landscape** and **Virtual Landscape Controls** from the View menu or click the 3D and 3D control icons on the toolbar.

- Click on the selected button in the control box to move across the landscape.

- To change the appearance of the virtual landscape select **Options** in the Tools

menu and click on **Virtual Landscape**. Here you can change lighting, sky colour, speed of movement, bearing and height data, height exaggeration and the keyboard shortcuts that can also be used to navigate over the landscape.

Downloading additional maps

You can choose from a wide range of maps from well-known cartographers such as Ordnance Survey and the French IGN. Maps can be downloaded as pre-defined bundles or self-defined, cut-your-own maps.

❶ From the Map menu on the Anquet Maps Toolbar, select an 'overview' type map such as the OS Road Map or OS GB Overview Map.

❷ Select **Map Manager** from the Tools menu or click the **D.Load** icon on toolbar (Anquet will connect to the MapSever and Map Manger will open).

❸ Choose the **Map Store** tab.

To select an Anquet Favourite Map Bundle (predefined bundles of map data):

- select 'Anquet's Favourite Map Bundles' from the drop-down list at the top of the **Map Manager**
- click on the '+' mark next to any series to see the bundles within
- click on the '+' mark next to any bundle to see the list of maps within
- click on each map in the bundle, the area it covers will be highlighted on the **Map** in the Anquet software.

To select a 'cut-your-own', or self-defined map:

- select the **Map Type** that you wish to obtain a price for from the drop-down list
- click on the **Define** button
- left-click on the map (press and hold the left mouse button), drag the mouse to dra a rectangle, and then release the left mouse button to finish drawing the rectangle
- click the **Add** button on the **Map Manager** – to redefine the area, simply click the **Redefine** button and drag the area out again as above
- click **Add**, you will see a price at the bottom of the Map Manager, for that area of mapping in the map type that you chose

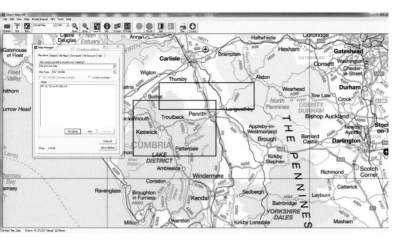

- You can define more areas – as you do so the displayed price will change.

❹ The price of the maps is shown at the bottom of the Map Manager – this is the price for all of the maps selected.

❺ Click **Add to Basket** and repeat and add any additional maps to basket.

❻ Select the **Basket** tab of the Map Manager to complete the transaction.

Buying digital mapping

Free or purchased digital mapping?

Most digital mapping is for all intents and purposes given away; the business model is to effectively make money selling the map data. However, there are some that are entirely free and you are going to learn how to use these first. Because they are free does not mean they are significantly inferior to purchased digital mapping. When you have become comfortable using these free programs you will be in a better position to know exactly what your individual requirements are and whether you wish to purchase digital mapping – *I personally recommend that you do*.

Satnav manufacturers' digital mapping

Most manufacturers support their satnavs with software that allows you to view and organise the maps you have bought from them, create routes, trails, tracks and waypoints and send them to/from your device. Some also let you geotag multimedia images and have the ability to transfer satellite images. These programs used to come on disks with the receiver but are now more often available as a download from the manufacturer's website.

Also some receiver makers offer web-based digital mapping which is really a hybrid of bespoke digital mapping and free-to-use web-based programs. These allow customers, who have purchased their receivers to plan, view and edit routes on their PC. These routes can then be downloaded onto their receiver and paper copies of the route and a route card printed off. Routes and tracks that have been created on the

↘ EXPERT TIPS

→ Avoid discs in shops, as who knows how long they have been sitting there. Download is safest – but check with the digital mapping manufacturer how old their maps are before you buy. The Ordnance Survey refreshes yearly – so your supplier should be able to give you up-to-date data.

→ GPX has now emerged as a standard for file interchange, and all software supports it, so you can always move. Don't get stuck with a supplier who won't allow you to extract your data or change provider.

receiver can also be uploaded onto the software and manipulated in the same way. However, they lack many of the features of bespoke digital mapping and you may have to pay for the privilege.

Many phone companies provide their own mapping software with specific features. Nokia's mobile phones provide Ovi mapping for free which can be used in conjunction with their digital mapping application **viNe**. Essentially viNe is a location-based journal where everything is tagged on your trip: photos, videos and even what music you listened to and where on your GNSS-enabled device.

When you have learned to use Google Earth and the digital mapping provided by the manufacturer of your device, you will have a better understanding of your requirements. You will very possibly also have identified a number of limitations of the systems you are using, and this knowledge will better inform your buying decisions. If you decide to purchase digital mapping these are the selection criteria I would suggest:

Compatibility and map availability

- Does my satnav work with the software? Go deeper than the promotional literature which says '*our software is compatible with Brunton, Eagle, Delorme, Garmin, Lowrance, Magellan*' etc.: check that your specific model can link and transfer/ receive data with the software. Don't be put off by your device not being listed, as the proliferation of devices means that software vendors simply can't check them all. However, most device manufacturers stick to set protocols for series of devices, so the simple check is to download the free software version and give it a try. With satnavs, compatibility is normally by manufacturer.
- Do you also want the software to work with a smartphone or PDA? Again check your specific model. It's also worth noting that many phones are sold under two or three different names by different networks. A quick Google search on the name of your phone may well reveal that it has a second and/or third name that you could look for. With smartphones, compatibility is normally by operating system – so if a system supports Windows Mobile 6.0, most likely it will run on all phones running that operating system.
- Can you buy topographical maps, at 1:25 000 and 1:50 000, of the area you wish to navigate? One of the major advantages of paid systems is the option to buy topographical maps that are not available for free. Check that the maps you are interested in are available within the software you are considering.

If the answer to any of these questions is 'No' move on to another piece of software or contact the company to ask if your requirements are likely to be met in the near future.

Minimum ordering quantities and price

It may be that you live and wish to navigate in only one part of a country, for example Scotland: check that you can buy a specific area of a map and do not have to buy one that covers the whole of the UK.

- Compare the prices of the same maps (scale and area covered): they can be surprisingly different.
- Are foreign maps available in areas you wish to travel to?
- Is satellite and aerial photography available?
- Are other map types important to you, such as waterways or flight charts?

Graphics

Can you create a 3D landscape using maps or aerial photographs which you can move through, following your tracks and routes, then print out? 3D maps are one of the best ways of learning to interpret topographical maps and I think an essential feature.

- Is your computer's graphics card powerful enough to handle this feature?

Computer requirement

Some digital mapping programs require a lot of computing power, especially if generating moving 3D images. In addition, map files, especially aerial photography, can be massive. Check that your computer has enough storage and will work with your operating system. Not just Apple Mac verses Microsoft Windows, but which version of Microsoft Windows? Again, most suppliers offer their software and trial maps for free, so you can simply download the software and give it a test drive.

User-friendliness

Since a great deal of the software is free to download, trial it first. Most manufacturers also supply free of charge a very small area of a map for you to do this. See if you like the interface, are you comfortable with the layout, does it work well on your computer.

While user-friendliness is very important, look beyond your first few days of perceived usage. Can you grow with this software? How are its advanced features for printing, and interacting with your data? Are simple things present like undo/redo?

Online service

Some manufacturers give access to additional facilities on their website:

- Can you upload your waypoints, routes, even photographs and share them with this software's user community online?
- Use their software with programs such as Google Earth free of change?
- Access national map agency mapping, such as Ordnance Survey 1:50 000 and 1:25 000, from any computer so when you are away from home you can plan and print routes?

Reliability and backup

Read user reviews on the internet and check manufacturers' claims, in particular:

- The software is reliable and not prone to 'freezing' or bugs?

- Software updates are easy to obtain and free.
- How long has the company been creating digital mapping systems? This can be an indication of how developed their software is, and how reliable the supplier is.
- Map updates – do you buy them at a discount?
- Technical backup – does the manufacturer's helpline have a good reputation?
- Is their website helpful, with FAQs and a Q&A section?
- If you download your maps, can you re-download them if you move computer or have a computer crash?

Satellite and aerial photography

This is what we have all become so used to with Google Earth and some digital mapping manufacturers also sell these images.

However, when buying satellite and aerial photography it is difficult to assess the quality of the images as different manufacturers use different terms to describe the image resolution (quality of the picture).

The two most used are Geometric Resolution (GR) or Spatial Resolution (SR) and this is how to compare the two:

- A GR of 1 m is equivalent to 1 pixels/m^2 in SR
- A GR of 15 cm is equivalent to 45 pixels/m^2 in SR.

Don't worry too much about the science, just the compare the numbers (GR is a portion of the earth's surface in a single pixel and SR defines the pixel size of an image representing the size of the surface area measured on the ground).

Another often held misconception is that 1 m GR aerial photography is twice as clear as 2 m GR aerial photography. This is wrong, it's actual 4 times! With 2 m GR, one computer pixel represents an area 2 m x 2 m on the ground. With 1 m GR, that same 2 m x 2 m area would be covered by 4 pixels.

Time to buy?

When you have considered all of these features weigh up the total cost, including all of the mapping/photography you are likely to require, and decide if it is worth it. Be prepared to spend as much on the digital mapping as you have on your satnav!

↘ WARNING

Always remember to include the local magnetic declination on maps you print as well as the grid (meridians) so you can give a grid reference, especially if you need help to get to you!

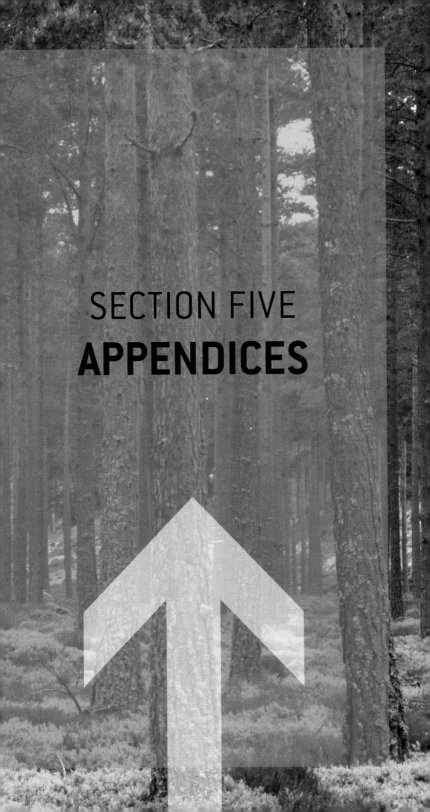

SECTION FIVE
APPENDICES

PLANNING, PREPARATION AND USEFUL TABLES

Care of your equipment

Four of Diamonds

People fall predominantly onto one side of their bodies and for this reason pack your spare items on the opposite side to their counterpart, items such as your compass and reading glasses. Ideally keep them on your person and not in your rucksack – in a fall you are much less likely to become separated from your jacket than a rucksack.

Lanyards

I attach lanyards to all essential items, even my gloves, and because of this I have devised a very straightforward system of being able to quickly attach and undo them from whatever I am fastening them to.

Using a bowline loop knot create a loop which you can slip the item easily through.

our compass, map case, GNSS and mobile telephone should always be securely attached on a lanyard to either your rucksack or ideally your jacket.

Compasses

The clear plastic of your compass scratches easily, so try to keep it in a pocket with nothing else. Always keep it away from your mobile phone, communications radio (SAR team members). If it gets dirty from mud and the like, wash it in running cold water only and dry with a soft cloth.

Satnav

Outdoor handheld satnavs are built to be rugged and weatherproof, nevertheless, the screens scratch easily and you should therefore use a clear screen protector. If the manufacturer does not provide these, they are readily available via the internet. Never use anything other than a damp cloth to wash your unit, solvents can seriously damage them. Most units have a rubber closure over the USB and external antenna slots, always make sure you securely close this after use to prevent the ingress of dust or water.

Binoculars

Always replace the lens caps when not in use and use a lint free cloth to clean the lenses when required.

Maps

If your maps are paper then a waterproof map holder/cover is needed.

Grid reference tool

Attach on the same lanyard as your compass, your slope angle and **Pacing and Timing** chart, if you have these, should also be securely attached to this lanyard.

Batteries

Keep spare batteries dry, ideally in a waterproof bag. For the choice and correct use see **Batteries** Section.

Mobile phones

If your phone is not waterproof, buy a small waterproof bag for it in clear plastic so you can see who is calling without removing it. If my journey is for leisure, I personally turn my phone off and enjoy the peace and beauty of my environment. If travelling over a period of days, I turn my phone at times I have pre-arranged times with my family/close friends. In either event, switch on the 'Save Battery' function if it has it: this will automatically turn the phone off when the batteries are low and leave enough power for it to make short calls when turned back on.

↘ EXPERT TIP

→ If you lose/damage your reading glasses use the compass' magnifying glass to read the map.

The twelve most common errors in navigation

ERROR	REMEDY
Lack of concentration.	Clear thinking is critical for accurate navigation. When calculating moves use the Brace Position to take you 'out' of the group. If navigating in difficult conditions tell the group you are not going to chat and instead concentrate only on navigation.
Making the map fit the environment.	When you think that you have your position located on the map, choose a feature on your map, predict where you are going to see it and then turn in this direction to see it. If it isn't there start your relocation procedure.
In-accurate compass bearings.	Always have the compass set in front of you and rote your body, not the compass to your objective. If taking a bearing from a map always use the brace position.
Compass Deviation	Check wristwatches, karabiners, ice axes, walking poles are all kept clear of your compass – even your jackets zip-pull!
Forgetting to correct for Magnetic Declination	Practice this so much that it becomes a conditioned reflex and you no longer have to think about it.
Walking the wrong way by 180°	When the Red light is on GO – a simple mnemonic to use to remember that it is the Red end of the magnetic needle points north. If the terrain is dangerous check your bearing with another member of your group.
Walking the wrong way by 90°	Make sure the orienting lines on your map are north/south when using the map to take bearings and NOT east/west.
Missing Environmental clues.	If in a group talk about your surroundings, this will keep environmental clues fresh in your mind. Alternatively write these down as you travel.

	This can easily happen with tracks and paths, streams, knowles and mountain ridges: any feature which is replicated in the land you are crossing. So if you cross a number of paths that are parallel to each other you can mistake your position because one of the paths is not marked on your map, you may have miscounted the number or be interpreting the map incorrectly.
Parallel Features. This is such a common error I have separately detailed how to avoid it in the techniques section.	In areas with similar features be aware of this risk and if the landscape features do not agree with what is on your map stop and if possible retrace your steps back to your last attack point. Otherwise start your relocation procedure.
Other group members seem unsure about the group's location.	Listen to the advice of others then make your own judgement!
Miscalculating distance travelled	Practice your Pacing and Timing in good weather.
Drift	High winds make it difficult to walk on a straight bearing so compensate for this.

Faulty compass?

Magnetic compasses can be influenced by many factors so you need to eliminate these before you determine that it is not functioning correctly.

When navigating under severe or difficult conditions a simple yet frequent mistake is to forget that the north of your compass needle is red – check you are using the **RED** for north.

If you are carrying a backup compass check the readings are the same on both and you still suspect their reliability continue to Step 4.

Check for metal objects close to it, such as a belt buckle, the under-wiring of a bra or is it near a wire fence? If so move it away from these objects.

Have you moved into an area with large deposits of iron ore, look at the geology of stones around you (*most iron ores are dark reddish brown*) and your map because these areas are often marked.

Having eliminated these possibilities:

1 Put your compass in your pocket and set your map using the simple celestial technique: the sun in the mornings will be eastwards and in the afternoons westwards – roughly orient your map north.

❷ Search for features in your surroundings; ideally look for linear features that you can align with your map, streams, rivers, walls, electricity transmission lines. Use environmental clues, which way is the prevailing wind, if you know this look at the trees and see which way they bend.

❸ Keeping your map set – remember your move around your map and not the other way round – take out your compass and place it on the map and see if the red needle north now matches your maps north (allowing for the local **Magnetic Declination**).

❹ If after following these steps you are sure that your compass is inoperative and you are not carrying a GNSS your best course of action is to stop your journey and return to where you started from if there is safety/shelter/your vehicle there, constantly checking your position on the map.

❺ If the terrain is not difficult, the weather good with a lot of daylight left, and you are confident reading the map continue to use the features, the sun and environmental clues to constantly reaffirm where you are and continue your journey.

Emergency calling procedure

I spent months researching this information and obtaining permission to publish it, dealing with government departments from the UK Cabinet Office to the US Department of Homeland Security, major mobile phone manufacturers, numerous mobile phone operators and many SAR organisations internationally: you need to know about them as one day it could save your life!

What do I need to know?

- In nearly all major economies over 90% of the population carry a mobile phone.
- More than 50% of all emergency calls in most of these countries are initiated from a mobile phone, in many this figure is over 80%.
- Mobile phones work well in towns and cities because network operators build the telephone masts in areas of dense population.
- Coverage is frequently over 90% of the population.
- It is very important to realise that this coverage is of the population and not of the land area, where the percentage is always much lower.

For example, the largest network operator in USA is Cingular TM with coverage of over 90% of the population, but their coverage and reception is limited in remote areas and small towns and the total land area they cover it is just over 50%.

In addition, geographic phone coverage varies by operators, such as Nextel TM, T-mobile TM and Verizon TM, as they each run their own phone masts in different locations and therefore all have different dead spots - areas of no coverage.

Because of this, governments, phone manufacturers and network operators have developed procedures and technology solutions to try and overcome these problems.

The essential knowledge

The emergency services numbers 112 and 911 are pre-programmed into all mobile phones or SIM cards. Dialling 112 anywhere in Europe and 911 anywhere in America, instead of another emergency number has major benefits. Networks give special priority to these emergency numbers and immediately initiate a special emergency set p to force the phone to make the call on any network, so even if your phone shows 'No ervice' or 'No Signal' it will still transmit if another network is available. Additionally most modern mobile phones can still dial these emergency numbers if:

The phone has no credit left
The phone is without a SIM card
An emergency number is entered instead of the PIN
The keypad is locked – so even if you do not have a phone but an injured person does and who may be unable to tell you the password, you can still call the emergency services.
The network is congested because of the volume of calls. This occurs when a major disaster such as an earthquake or flooding occurs and too many people are using the network.
In the US a deactivated cell phone will still complete a 911 call

The 'wise to know' knowledge

These facts will help you get more from your phone yet it is not essential; you may want to jump straight to 'What do I need to tell the person who takes my call?'

All network operators can provide the emergency services with the telephone number of the originator of a wireless 911/112 call and the location of the cell site or base station transmitting the call. The accuracy of this varies considerably depending upon the number of masts and their proximity to you from a few hundred metres to several kilometres from the actual site. The time taken to do this varies considerably from country to country. In the USA by law it is within six minutes of a valid request, in the UK it is usually hours and some countries can be days. (This information can also be obtained when you are not making a call).

The USA are the first country in the world to implement an enhanced positioning service called E911, providing more precise location information, specifically, the latitude and longitude of the caller which must be accurate to within 50 to 300 meters depending on the type of technology used, usually the phone's inbuilt GNSS and again within six minutes of a valid request. This implementation should be complete in 2012

Some countries, such as Finland, recommend that you carry a mobile phone even when beyond cellular coverage, in remote areas as the radio signal of a mobile phone attempting to connect to a base station can be detected by overflying rescue aircraft with special detection gear. When a phone is in 'idle mode' (not during active calls), every time the phone attaches to the network it will be provided with a Location Update timer value that it uses to contact the network, usually every hour. When there is no base station (no network can be found), the handset will search for the network to lock onto, so again it will be transmitting from time to time; the timing/frequency of this search varies from handset to handset.

During major emergencies cellular networks can experience congestion due to increased call volumes and/or damage to network facilities, severely curtailing the ability of emergency services and national security personnel to make emergency calls. With an increasing number of these professionals relying on cell phones various governments have introduced different schemes to provide priority for emergency calls made from cellular telephones.

In the UK the scheme that allows the mobile telephone networks to restrict access in a specific area to registered numbers – specific phones – in a major disaster and override these logjams is called MTPAS (Mobile Telecommunication Privileged Access

Scheme), which replaced ACCOLC (Access Overload Control) in 2009. There are 16 levels of access controlled and the levels 10 and 11 are designated for 112/999 calls and the likelihood of not being able to get through on these numbers is very remote.

In the USA a similar system is in place called WPS (Wireless Priority Service). Yet when you dial the 911 emergency number your mobile phone will ignore this restriction and always proceed to initiate the call.

In extreme emergencies, such as major terrorist attacks, total network has been implemented. In the event of major terrorist attacks or assault by foreign forces government's can order network operators to lockdown all mobile phone access. During these periods it may still be possible to send a text message (SMS) on some phones, in particular smartphones with Wi-Fi/LAN access.

Satellite phones are not subject to these constraints.

- So check with your network operator what is available in your country and if travelling abroad, do the same before you depart.

What do I need to tell the person who takes my call in an emergency?
Prior to calling, collate this information to give to the emergency services and if possible write these details down to conserve your mobile phone's battery and your memory!

- Your name
- Your exact position on the map, ideally a grid reference and map number or your last known location, which direction you have been travelling and estimated time from it – If you are lost write down features you can see around and where they are. Rescue teams will help you locate your position from the description of your surroundings, shortening a potential protracted search, and getting assistance to you quicker.
- Other telephone numbers in the group. There is no need to give them yours because even if you have 'Number Withheld' the network will still provide the operator with your number.
- Nature of the incident.
- Time of the incident.
- Number of people in the group.
- Names and injuries of any casualties.
- Weather conditions
- Equipment at the site – such as First aid kits, warm clothing, shelter, flares etc. Any distinguishing land feature or marker at the accident site.

What do I need to do?
To make a call to the emergency services follow these procedures. If one stage fails move on to the next.

STAGE I
- Stand still
- Dial the emergency number
- Hold the handset next to your ear and protect the microphone from the wind
- Wait for up to one minute for a connection.
- Keep the phone held in the same position to maintain the connection.
- When you have made contact with the emergency services establish a calling schedule, say every 20 mins, and turn off the phone in-between times.

Important: If you had 'No Service' or 'No Signal' on your mobile phone only you can contact 112/911 – they cannot phone you back.

STAGE II

Redial the emergency number standing in the same position but hold the handset next to your other ear – it could be that your head is in-between the only mobile phone mast and your phone.

STAGE III

Move to higher ground or around an obstacle you think may be blocking your signal to a mobile phone mast, if it is safe to do so. When you reach this position, stand still and repeat Stages I and II if necessary.

STAGE IV

Send a text message (SMS) to the emergency operator – you must have preregistered before sending (see **micronavigation.org/112**) – stating your location, the nature of the emergency and what assistance you require.

SMS text messages can also be sent to landlines/fixed telephones in some countries. The recipient is telephoned on your behalf and your message is automatically read out loud to them. You should test this with your designated ICE number (see **ICE**, below) before you venture out.

STAGE V

If you are unable to make any contact whatsoever using your mobile phone, you should keep it on (see Finland) and then use the **International Distress Signal**.

- Six quick successive whistle blasts (*if you do not have a whistle or cannot whistle, six quick successive torch/camera flashes, or wave any available bright clothing*).
- Wait one minute.
- Repeat the signal every five minutes.
- When you get a reply it is three blasts or flashes within one minute
- Keep repeating the signal until help arrives. This is important and helps those coming to you to actually find you.
- If you are carrying flares a red flare is internationally accepted as a distress signal.

Important note: In the USA and Canada the Distress Signal is three whistle blasts.

Ground-to-air emergency signals

If the emergency services have started a search for you, either after you have made contact or if your party has been reported missing, in remote areas they will often call in an air asset to assist with their search, usually a SAR helicopter.

A difficulty frequently encountered by the aircrew of these aircraft is that members of the public often harmlessly wave at low flying aircraft, so knowing how to communicate with them correctly will greatly assist both them and you. There are two methods of transmitting emergency messages:

Emergency Signals – The body can be used to transmit messages. The individual stands in an open area to make the signals and ensures that the background (*as seen from the air*) is not confusing, goes through the motions slowly, and repeats each signal until it has been understood.

Need medical assistance

Do not attempt to land here

Land here

Yes

Ground-to-air emergency signals

No

All OK – do not wait

Pick us up

- If the aircraft is a helicopter the aircrew will usually acknowledge your message with a thumbs up.
- If it is a fixed wing aircraft the pilot indicates that ground signals have been understood by rocking the wings laterally.

In Case of Emergency (ICE)

ICE stands for 'In Case of Emergency' and is what first responders, such as paramedics, fire-fighters, police officers and SAR team members will look for on your mobile phone if you are involved in an accident.

Getting started
On most mobile phones you simply need to select 'Contacts' and choose 'Add New Contact', then enter the letters 'ICE' next to the name, followed by the telephone number of your next of kin. Enter daytime and evening numbers where this is possible. It is also a good place to include mention of your blood type and any allergies.

What should I do next?
Make sure the person whose name and number you are giving has agreed to be your 'ICE partner'. You should also make sure they know about any medical conditions that could affect your emergency treatment, including allergies and any medication you are on. If you are under 18, your ICE partner should be your parents or Guardian, as other people will not be able to make decisions for you if you are admitted to hospital.

→ Keep your mobile phone dry in a plastic bag and somewhere warm and where the ringer can be heard.

→ Carry a spare phone battery and make sure the battery in the phone is fully charged before you depart.

→ In an emergency only use the phone to talk to the emergency services as connection time uses a lot of battery. Have all the details to hand before phoning the emergency services; write them down if possible Grid reference plus any injuries and names of casualties.

→ When you have made contact with the emergency services establish a calling schedule, say every 20 mins, and turn off the phone in-between times. However if you had 'No Service' and 'No Signal' only you can contact 112/911 – they cannot phone you.

First responders

Search the casualty's mobile phone contacts list for ICE. If the phone is locked try pressing the * key 3 times to retrieve the information. As mobile phones are standardised this will allow the ICE information to be read even if the phone is locked.

Location/Position from a mobile

All mobile/cellular network operators can request location updates from a given handset and calculate your location from the towers receiving your signal: this is usually accurate in built-up areas – yet in remote areas, where the cells are huge, it may not be very accurate with errors in kilometres. This process can take many hours, even days, to obtain in some countries.

In America cellular/mobile location is a legal requirement under the FCC 911 Mandate. All 911 calls must be relayed to a call centre, regardless of whether the mobile phone user is a customer of the network being used and network operators must identify the phone number and cell phone tower plus the latitude and longitude of callers within 50–300 meters, within six minutes of a request. If the phone has Assisted GPS the location will be the most accurate.

At times of network congestion, location updates may be turned off to relieve processor load on the MSC so do not rely upon your location being given in this way.

A mobile phone is never a replacement for common sense, good navigational skills, having the right kit and an ability to know when to turn back!

In some countries, such as Finland, the rescue services suggest hikers carry mobile phones in case of emergency even when deep in the forests beyond cellular coverage, as the radio signal of a mobile phone attempting to connect to a base station can be detected by overflying rescue aircraft with special detection gear.

SMS Text Messages can be sent to landlines/fixed telephones in some countries. The recipient is telephoned on your behalf and your message is automatically read out loud to them.

A traveller visiting a foreign country does not have to know the local emergency numbers as any of the above numbers will work in any country.

In the UK the scheme that allows the mobile telephone networks to restrict access in a specific area to registered numbers in a major disaster is called MTPAS (Mobile Telecommunication Privileged Access Scheme) which replaced ACCOLC (Access Overload Control) in 2009. In the USA a similar system is in place called WPS (Wireless Priority Service). Yet when you dial an emergency number your mobile phone will ignore this restriction and always proceed to initiate the call.

INDEX